高 等 学 校 教 材

Introduction to Materials Science and Engineering
材料科学与工程导论（双语）

陈克正　王　玮　刘春廷　等编

化学工业出版社
·北京·

本教材是材料科学与工程导论的双语教材，以现行"材料科学与工程导论"课程标准为依据，结合中文教材，以国外原版教材做参考并根据国内的教学情况及材料科学研究的最新进展对教材内容进行适度的整合。全书共分9章，具体内容包括：绪论，固体材料的结构，常用工程材料（高分子材料、金属材料、陶瓷材料和复合材料）的结构、力学性能、成分、加工工艺以及应用前景，常用工程材料的化学性能（耐腐蚀性能）和物理性能（电、磁、热和光学性能）以及新型材料（生物材料、纳米材料和智能材料）的介绍等内容。

本教材可供大专院校材料科学与工程及相关专业师生使用，也可供从事材料科学与工程研究、开发及管理的人员参考。

图书在版编目（CIP）数据

材料科学与工程导论（双语）/陈克正等编．—北京：化学工业出版社，2011.7（2024.1重印）
高等学校教材
ISBN 978-7-122-11758-8

Ⅰ. 材⋯ Ⅱ. 陈⋯ Ⅲ. 材料科学-双语教学-高等学校-教材 Ⅳ.TB3

中国版本图书馆CIP数据核字（2011）第131568号

责任编辑：杨　菁　满悦芝　　　　　　文字编辑：颜克俭
责任校对：边　涛　　　　　　　　　　装帧设计：杨　北

出版发行：化学工业出版社（北京市东城区青年湖南街13号　邮政编码100011）
印　　装：涿州市般润文化传播有限公司
787mm×1092mm　1/16　印张14¼　字数360千字　2024年1月北京第1版第7次印刷

购书咨询：010-64518888　　　　　　　售后服务：010-64518899
网　　址：http://www.cip.com.cn

凡购买本书，如有缺损质量问题，本社销售中心负责调换。

定　价：39.00元　　　　　　　　　　　　　　　　　　版权所有　违者必究

前　　言

进入21世纪，新材料、信息和生物技术已成为科学研究和新兴战略产业发展的最重要领域。为了适应材料科学与技术的发展，培养学生跟踪国际材料科学与技术发展前沿的能力，使学生成为材料科学领域的创新型人才，国内很多高校都面向大学本科学生开设了"材料科学与工程导论"双语课程作为材料学科的基础课。

"材料科学与工程导论"双语课程是我校为材料类宽专业本科学生开设的专业课程，也是我校最早开设的双语课程，自2002年起，每年有五百余人修读本课程。2008年该课程成为山东省双语教学示范课程，2009年获得国家双语教学示范课程建设项目的资助。在"材料科学与工程导论"双语课程的建设中，教材建设是该课程建设的重中之重，也是双语教学必需解决的难题之一。"材料科学与工程导论"双语教材的编写是适应高等教育国际化进程、培养具有国际竞争力的材料领域高素质人才的重要一环。

目前，国外现有的原版英文教材不能全面满足我国材料类各专业大学本科阶段学生"材料科学与工程导论"专业基础课程双语教学的需要，也不能适应宽口径材料专业人才培养模式的需要。现阶段本科生不具备使用多本英文原版教材进行课程学习的基础和条件。因此，我们结合我校材料类专业设置特点（设高分子材料科学与工程、材料物理、材料化学、无机非金属材料、金属材料五个专业）、课程定位（专业基础必修课程）和学生学业和职业发展的需要，以现行"材料科学与工程导论"课程标准为依据，结合中文教材，以国外原版教材做参考并根据国内的实际教学情况及材料科学研究的最新进展对教材内容进行适度的整合，力图编写出适合我国大学本科生的"材料科学与工程导论"双语教材。

全书共分9章，具体内容包括：绪论，固体材料的结构，常用工程材料（高分子材料、金属材料、陶瓷材料和复合材料）的结构、力学性能、成分、加工工艺以及应用前景，常用工程材料的化学性能（耐腐蚀性能）和物理性能（电、磁、热和光学性能）以及新型材料（生物材料、纳米材料和智能材料）的介绍等内容。

本书的编写者为青岛科技大学陈克正教授（第9章），青岛科技大学王玮副教授（第2、3、6章），青岛科技大学刘春廷老师（第4、5、7章），青岛科技大学徐磊老师（第1、8章）。全书由陈克正教授任主编。

本书的编写得到了国家双语教学示范课程建设项目、青岛科技大学双语教学示范课程建设项目的资助。青岛科技大学各级领导对本书的编写和出版十分关心和支持，编者再次一并表示衷心的感谢。

由于编者水平有限，经验不足，书中难免有不足之处，恳请专家和广大读者批评指正。

编者
2011年6月

Contents

Chapter 1　Introduction ··· 1
　Learning Objectives ··· 1
　1.1　Historical Perspective ··· 1
　1.2　What is Materials Science and Engineering? ······················· 2
　1.3　Why Study Materials Science and Engineering? ·················· 5
　1.4　Classification of Materials ·· 5
　1.5　Advanced Materials ··· 9
　1.6　Modern Materials' Needs ··· 10
　References ··· 11

Chapter 2　The Structure of Crystalline Solids ··························· 13
　Learning Objectives ··· 13
　2.1　Atomic Structure and Interatomic Bonding ······················· 13
　　2.1.1　Fundamental Concepts ··· 14
　　2.1.2　Bonding Forces and Energies ···································· 14
　　2.1.3　Atomic Bonding in Solids ··· 16
　2.2　Crystal Structures ·· 22
　　2.2.1　Fundamental Concepts ··· 22
　　2.2.2　Metallic Crystal Structures and Crystal Systems ········ 23
　　2.2.3　Crystallographic Points, Directions, and Planes ········ 30
　　2.2.4　Crystalline and Noncrystalline Materials ·················· 37
　2.3　Imperfections in Solids ·· 40
　　2.3.1　Point Defects in Metals ·· 40
　　2.3.2　Dislocations—Linear Defects ···································· 43
　　2.3.3　Interfacial Defects ·· 44
　　2.3.4　Bulk or Volume Defects ··· 46
　References ··· 48

Chapter 3　Polymer Materials ·· 49
　Learning Objectives ··· 49
　3.1　Polymer Structures ·· 49
　　3.1.1　Introduction ·· 49
　　3.1.2　Fundamental Concepts ··· 49
　　3.1.3　Polymer Molecules ·· 50
　　3.1.4　Designation of Polymers ·· 50
　　3.1.5　Commonly Used Polymers ······································· 51
　　3.1.6　The Chemistry of Polymer Molecules ······················· 58
　3.2　Crystallization, Melting and Glass Transition Phenomena in Polymers ··············· 65

3.3　Mechanical Properties of Polymers ·· 66
　　3.3.1　Stress-Strain Behavior ··· 67
　　3.3.2　Macroscopic Deformation ··· 68
　　3.3.3　Viscoelastic Deformation ··· 70
3.4　Polymer Types ··· 71
　　3.4.1　Plastics ·· 71
　　3.4.2　Elastomers ··· 72
　　3.4.3　Fibers ··· 73
　　3.4.4　Miscellaneous Applications Coatings ··· 73
3.5　Processing of Polymers ··· 74
　　3.5.1　Polymerization ··· 75
　　3.5.2　Polymer Additives ·· 76
　　3.5.3　Forming Techniques for Plastics ·· 78
　　3.5.4　Fabrication of Elastomers ·· 80
　　3.5.5　Fabrication of Fibers and Films ··· 81
References ·· 84

Chapter 4　Metallic Materials ·· 85
Learning Objectives ·· 85
4.1　Mechanical Properties of Metals ·· 85
　　4.1.1　Introduction ··· 85
　　4.1.2　Tensile Test ··· 86
　　4.1.3　Hardness Testing ·· 90
4.2　Dislocations and Strengthening ·· 91
　　4.2.1　The Role of Dislocations ·· 91
　　4.2.2　Work Hardening ··· 93
　　4.2.3　Grain Size Strengthening ·· 93
　　4.2.4　Alloy Hardening ··· 94
4.3　Failure ··· 96
　　4.3.1　Introduction ··· 96
　　4.3.2　Fundamentals of Fracture ··· 97
　　4.3.3　Ductile Fracture ·· 97
　　4.3.4　Brittle Fracture ··· 98
　　4.3.5　Fracture Mechanics in Design ·· 99
　　4.3.6　Fracture Toughness ··· 100
　　4.3.7　Fatigue ·· 101
　　4.3.8　Creep ·· 101
4.4　Phase Diagrams and Phase Transformations in Metals ····································· 102
　　4.4.1　Introduction ·· 102
　　4.4.2　Phase Diagrams ··· 103
　　4.4.3　Phase Transformations ·· 107
4.5　Applications and Processing of Metal Alloys ·· 108
　　4.5.1　Introduction ·· 108

  4.5.2 Types of Metal Alloys ······ 109
  4.5.3 Fabrication of Metals ······ 113
  4.5.4 Thermal Processing of Metals ······ 115
 References ······ 120

Chapter 5 Ceramic Materials ······ 123
 Learning Objectives ······ 123
 5.1 Structures and Properties of Ceramics ······ 123
  5.1.1 Introduction ······ 123
  5.1.2 Ceramic Structures ······ 123
  5.1.3 Mechanical Properties of Ceramics ······ 127
 5.2 Application and Processing of Ceramics ······ 129
  5.2.1 Types and Applications of Ceramics ······ 129
  5.2.2 Fabrication and Processing of Ceramics ······ 131
 References ······ 136

Chapter 6 Composite Materials ······ 137
 Learning Objectives ······ 137
 6.1 Introduction ······ 137
 6.2 Particle-Reinforced Composites ······ 138
  6.2.1 Large-Particle Composites ······ 139
  6.2.2 Dispersion-Strengthened Composites ······ 140
 6.3 Fiber-Reinforced Composites ······ 140
  6.3.1 The Fiber Phase ······ 140
  6.3.2 The Matrix Phase ······ 141
 6.4 Polymer-Matrix Composites ······ 142
  6.4.1 Glass Fiber-Reinforced Polymer (GFRP) Composites ······ 142
  6.4.2 Carbon Fiber-Reinforced Polymer (CFRP) Composites ······ 143
  6.4.3 Aramid Fiber-Reinforced Polymer Composites ······ 143
 6.5 Metal-Matrix Composites ······ 145
 6.6 Ceramic-Matrix Composites ······ 146
 References ······ 149

Chapter 7 Corrosion and Degradation of Materials ······ 150
 Learning Objectives ······ 150
 7.1 Introduction ······ 150
 7.2 Corrosion of Metals ······ 150
  7.2.1 Electrochemical Considerations ······ 151
  7.2.2 Corrosion Rates ······ 153
  7.2.3 Passivity ······ 153
  7.2.4 Environmental Effects ······ 153
  7.2.5 Forms of Corrosion ······ 154

		7.2.6 Corrosion Environments	158
		7.2.7 Corrosion Prevention	159
7.3	Corrosion of Ceramic Materials		159
7.4	Degradation of Polymers		160
	7.4.1	Swelling and Dissolution	160
	7.4.2	Bond Rupture	160
	7.4.3	Weathering	161
	References		163

Chapter 8 Electrical/Thermal/Magnetic/Optical Properties of Materials 164

Learning Objectives 164

- 8.1 Introduction 165
- 8.2 Electrical Properties of Materials 165
  - 8.2.1 Metals and Alloys 166
  - 8.2.2 Semiconductors 167
  - 8.2.3 Ionic Ceramics and Polymers 170
- 8.3 Thermal Properties of Materials 170
  - 8.3.1 Heat Capacity 170
  - 8.3.2 Thermal Expansion 171
  - 8.3.3 Thermal Conductivity 172
  - 8.3.4 Thermal Stresses 172
- 8.4 Magnetic Properties of Materials 174
  - 8.4.1 Diamagnetism, Paramagnetism and Ferromagnetism 174
  - 8.4.2 Antiferromagnetism and Ferrimagnetism 177
  - 8.4.3 The Influence of Temperature on Magnetic Behavior 179
  - 8.4.4 Domains, Hysteresis and Magnetic Anisotropy 179
  - 8.4.5 Superconductivity 181
- 8.5 Optical Properties of Materials 186
  - 8.5.1 Interaction of Light with Matter 186
  - 8.5.2 Atomic and Electronic Interactions 187
  - 8.5.3 Refraction, Reflection, Absorption and Transmission 190
  - 8.5.4 Opacity and Translucency in Insulators 192
  - 8.5.5 Applications of Optical Phenomena 192

References 197

Chapter 9 Biomaterials/Nanomaterials/Smart Materials 198

Learning Objectives 198

- 9.1 Biomaterials 198
  - 9.1.1 Definition of Biomaterials 199
  - 9.1.2 Performance of Biomaterials 202
  - 9.1.3 Brief Historical Background 203
- 9.2 Nanotechnology and Nanomaterials 205

9.2.1	Introduction	205
9.2.2	Examples of Current Achievements and Paradigm Shifts	209
9.3	Smart Materials	214
9.3.1	Introduction	214
9.3.2	Shape Memory Alloys	215
9.3.3	Applications of Smart Materials	219
Reference		220

Chapter 1　Introduction

Learning Objectives
After careful study of this chapter you should be able to do the following:
1. *List six different property classifications of materials that determine their applicability.*
2. *Cite the four components that are involved in the design, production, and utilization of materials, and briefly describe the interrelationships between these components.*
3. *Cite three criteria that are important in the materials selection process.*
4. *(a) List the three primary classifications of solid materials, and then cite the distinctive chemical feature of each.*
   (b) Note the two types of advanced materials and, for each, its distinctive feature (s).

1.1　Historical Perspective

　　The designation of successive historical epochs as the Stone, Copper, Bronze, and Iron Ages reflects the importance of materials to mankind. Human destiny and materials resources have been inextricably intertwined since the dawn of history; however, the association of a given material with the age or era that it defines is not only limited to antiquity. The present nuclear and information ages owe their existences to the exploitation of two remarkable elements, uranium and silicon, respectively. Even though modern materials ages are extremely time compressed relative to the ancient metal ages they share a number of common attributes. For one thing, these ages tended to define sharply the material limits of human existence. Stone, copper, bronze, and iron meant successively higher standards of living through new or improved agricultural tools, food vessels, and weapons. Passage from one age to another was (and is) frequently accompanied by revolutionary, rather than evolutionary, changes in technological endeavors.

　　It is instructive to appreciate some additional characteristics and implications of these materials ages. For example, imagine that time is frozen at 1500 BC and we focus on the Middle East, perhaps the world's most intensively excavated region with respect to archaeological remains. In Asia Minor (Turkey) the ancient Hittites were already experimenting with iron, while close by to the east in Mesopotamia (Iraq), the Bronze Age was in flower. To the immediate north in Europe, the south in Palestine, and the west in Egypt, peoples were enjoying the benefits of the Copper and Early Bronze Ages. Halfway around the world to the east, the Chinese had already melted iron and demonstrated a remarkable genius for bronze, a copper—tin alloy that is stronger and easier to cast than pure copper. Further to the west on the Iberian Peninsula (Spain and Portugal), the Chalcolithic period, an overlapping Stone and Copper Age held sway, and in North Africa survivals of the Late Stone

Age were in evidence. Across the Atlantic Ocean the peoples of the Americas had not yet discovered bronze, but like others around the globe, they fashioned beautiful work in gold, silver, and copper, which were found in nature in the free state (i. e., not combined in oxide, sulfide, or other ores).

Why materials resources and the skills to work them were so inequitably distributed cannot be addressed here. Clearly, very little technological information diffused or was shared among peoples. Actually, it could not have been otherwise because the working of metals (as well as ceramics) was very much an art that was limited not only by availability of resources, but also by cultural forces. It was indeed a tragedy for the Native Americans, still in the Stone Age three millennia later, when the white man arrived from Europe armed with steel (a hard, strong iron-carbon alloy) guns. These were too much of a match for the inferior stone, wood, and copper weapons arrayed against them. Conquest, colonization, and settlement were inevitable. And similar events have occurred elsewhere, at other times, throughout the world. Political expansion, commerce, and wars were frequently driven by the desire to control and exploit materials resources, and these continue unabated to the present day.

When the 20th century dawned the number of different materials controllably exploited had, surprisingly, not grown much beyond what was available 2000 years earlier. A notable exception was steel, which ushered in the Machine Age and revolutionized many facets of life. But then a period ensued in which there was an explosive increase in our understanding of the fundamental nature of materials. The result was the emergence of polymeric (plastic), nuclear, and electronic materials, new roles for metals and ceramics, and the development of reliable ways to process and manufacture useful products from them. Collectively, this modern Age of Materials has permeated the entire world and dwarfed the impact of previous ages.

Only two representative examples of a greater number scattered throughout the book will underscore the magnitude of advances made in materials within a historical context. In Figure 1.1 the progress made in increasing the strength-to-density (or weight) ratio of materials is charted. Two implications of these advances have been improved aircraft design and energy savings in transportation systems. Less visible but no less significant improvements made in abrasive and cutting tool materials are shown in Figure 1.2. The 100-fold tool cutting speed increase in this century has resulted in efficient machining and manufacturing processes that enable an abundance of goods to be produced at low cost. Together with the dramatic political and social changes in Asia and Europe and the emergence of interconnected global economies, the prospects are excellent that more people will enjoy the fruits of the earth's materials resources than at any other time in history.

1.2 What is Materials Science and Engineering?

Materials science and engineering (MSE) is an interdisciplinary field concerned with inventing new materials and improving previously known materials by developing a deeper understanding of microstructure—composition—synthesis— processing relationships. The term

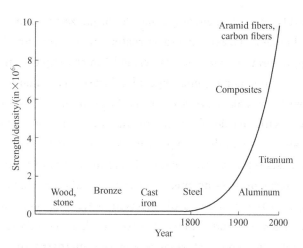

Figure 1.1 Chronological advances in the strength-to-density ratio of materials. Optimum safe load-bearing capacities of structures depend on the strength-to-density ratio. The emergence of aluminum and titanium alloys and, importantly, composites is responsible for the dramatic increase in the 20th century

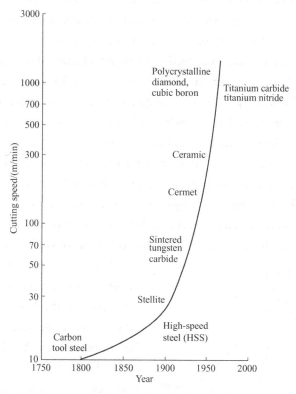

Figure 1.2 Increase in machining speed with the development over time of the indicated cutting tool materials

composition means the chemical make-up of a material. The term structure is at this point a nebulous term that deserves some explanation, as seen at different levels of detail. The structure of a material usually relates to the arrangement of its internal components. Subatomic structure involves electrons within the individual atoms and interactions with their nuclei. On an atomic level, structure encompasses the organization of atoms or molecules relative to one another. The next larger structural realm, which contains large groups of atoms that are normally agglomerated together, is termed "microscopic", meaning that which is subject to direct observation using some type of microscope. Finally, structural elements that may be viewed with the naked eye are termed "macroscopic." Materials science and engineering not only deal with the development of materials, but also with the synthesis and processing of materials and manufacturing processes related to the production of components. The term synthesis refers to how materials are made from naturally occurring or man-made chemicals. The term processing means how materials are shaped into useful components to cause changes in the properties of different materials. One of the most important functions of materials scientists and engineers is to establish the relationships between a material or a device's properties and performance and the microstructure of that material, its composition, and the way the material or the device was synthesized and processed. In materials science, the emphasis is on the underlying relationships between the synthesis and processing, structure and properties of materials. In materials engineering, the focus is on how to translate or transform materials into a useful device or structure.

One of the most fascinating aspects of materials science involves the investigation of a material's structure. The structure of materials has a profound influence on many properties of materials, even if the overall composition does not change! For example, if you take a pure copper wire and bend it repeatedly, the wire not only becomes harder but also become increasingly brittle! Eventually, the pure copper wire becomes so harder and brittle that it will break! The electrical resistivity of wire will also increase as we bend it repeatedly. In this simple example, take note that we did not change the material's composition (i.e., its chemical make up). The changes in the material's properties are due to a change in its internal structure. If you look at the wire after bending, it will look the same as before; however, its structure has been changed at a very small or microscopic scale. The structure at this microscopic scale is known as microstructure. If we can understand what has changed microscopically, we can begin to discover ways to control the material's properties.

Let's examine an example using the materials science and engineering tetrahedron presented on the Figure 1.3. Steels, as you may know, have been used in manufacturing for more than a hundred years, but they probably existed in a crude form during the Iron Age, thousand of years ago. In the manufacture of automobile chassis, a material is needed that possessed extremely high strength but is easily formed into aerodynamic contours. Another consideration is fuel-efficiency, so the sheet steel must also be thin and lightweight. The sheet steels should also

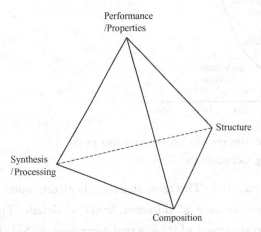

Figure 1.3 Scope of materials science and engineering

be able to absorb significant amounts of energy in the event of a crash, thereby increasing vehicle safety. These are somewhat contradictory requirements.

Thus, in this case, materials scientists are concerned with the sheet steel's ①composition; ②strength; ③weight; ④energy absorption properties; and⑤malleability (formability).

Materials scientists would examine steel at a microscopic level to determine if its properties can be altered to meet all of these requirements. They also would have to consider the cost of processing this steel along with other considerations. How can we shape such steel into a car chassis in a cost-effective way? Will the shaping process itself affect the mechanical properties of the steel? What kind of coatings can be developed to make the steel corrosion resistant? In some applications, we need to know if these steels could be welded easily. From this discussion, you can see that many issues need to be considered during the design and materials selection for any product.

1.3 Why Study Materials Science and Engineering?

Why do we study materials? Many an applied scientist or engineer, whether mechanical, civil, chemical, or electrical, will at one time or another be exposed to a design problem involving materials. Examples might include a transmission gear, the superstructure for a building, an oil refinery component, or an integrated circuit chip. Of course, materials scientists and engineers are specialists who are totally involved in the investigation and design of materials.

Many times, a materials problem is one of selecting the right material from the many thousands that are available. There are several criteria on which the final decision is normally based. First of all, the in-service conditions must be characterized, for these will dictate the properties required of the material. On only rare occasions does a material possess the maximum or ideal combination of properties. Thus, it may be necessary to trade off one characteristic for another. The classic example involves strength and ductility; normally, a material having a high strength will have only a limited ductility. In such cases a reasonable compromise between two or more properties may be necessary.

A second selection consideration is any deterioration of material properties that may occur during service operation. For example, significant reductions in mechanical strength may result from exposure to elevated temperatures or corrosive environments.

Finally, probably the overriding consideration is that of economics: What will the finished product cost? A material may be found that has the ideal set of properties but is prohibitively expensive. Here again, some compromise is inevitable. The cost of a finished piece also includes any expense incurred during fabrication to produce the desired shape. The more familiar an engineer or scientist is with the various characteristics and structure—property relationships, as well as processing techniques of materials, the more proficient and confident he or she will be to make judicious materials choices based on these criteria.

1.4 Classification of Materials

Solid materials have been conveniently grouped into three basic classifications: polymers, ceramics, and metals. This scheme is based primarily on chemical makeup and atomic structure, and most materials fall into one distinct grouping or another, although there are some intermediates. In addition, there are the composites, combinations of two or more of the above three basic material classes. A brief explanation of these material types and representative characteristics is offered next. Another classification is advanced materials—those used in high-technology applications—viz. semiconductors and biomaterials materials; these are discussed in Section 1.5.

Polymers

Polymers include the familiar plastic and rubber materials. Many of them are organic compounds that are chemically based on carbon, hydrogen, and other nonmetallic elements (O, N, and Si). Furthermore, they have very large molecular structures, often chain-like in nature, which have a backbone of carbon atoms. Some of the common and familiar polymers are polyethylene (PE), nylon, poly (vinyl chloride) (PVC), polycarbonate (PC), polystyrene (PS), and silicone rubber. These materials typically have low densities (Figure 1.4), whereas their mechanical characteristics are generally dissimilar to the metallic and ceramic materials—they are not as stiff nor as strong as these other material types (Figures 1.5 and Figure 1.6). However, on the basis of their low densities, many times their stiffnesses and strengths on a per mass basis are comparable to the metals and ceramics. In addition, many of the polymers are extremely ductile and pliable (i.e., plastic), which means they are easily formed into complex shapes. In general, they are relatively inert chemically and unreactive in a large number of environments. One major drawback to the polymers is their tendency to soften and/or decompose at modest temperatures, which, in some instances, limits their use. Furthermore, they have low electrical conductivities (Figure 1.8) and are nonmagnetic. Chapter 3 is devoted to discussions of the structures, properties, applications, and processing of polymeric materials.

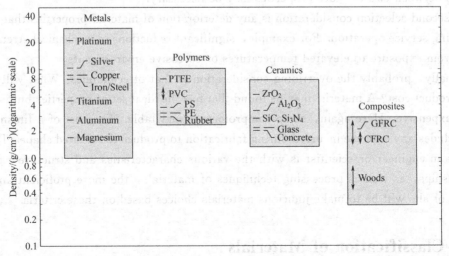

Figure 1.4 Bar-chart of room temperature density values for various metals, ceramics, polymers, and composite materials

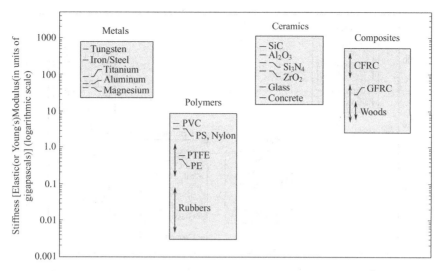

Figure 1.5 Bar-chart of room temperature stiffness (i. e., elastic modulus) values for various metals, ceramics, polymers, and composite materials

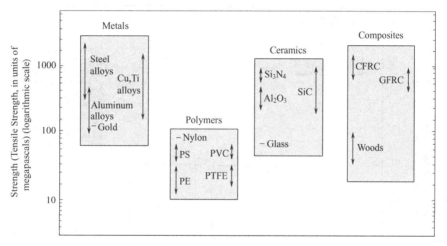

Figure 1.6 Bar-chart of room temperature strength (i. e., tensile strength) values for various metals, ceramics, polymers, and composite materials

Metals

Materials in this group are composed of one or more metallic elements (such as iron, aluminum, copper, titanium, gold, and nickel), and often also nonmetallic elements (for example, carbon, nitrogen, and oxygen) in relatively small amounts. Atoms in metals and their alloys are arranged in a very orderly manner (as discussed in Chapter 2), and in comparison to the ceramics and polymers, are relatively dense (Figure 1.4). With regard to mechanical characteristics, these materials are relatively stiff (Figure 1.5) and strong (Figure 1.6), yet are ductile (i. e., capable of large amounts of deformation without fracture), and are resistant to fracture (Figure 1.7), which accounts for their widespread use in structural applications. Metallic materials have large numbers of nonlocalized electrons; that is, these electrons are not bound to particular atoms. Many properties of metals are directly attributable to these electrons. For example, metals are extremely good conductors of electric-

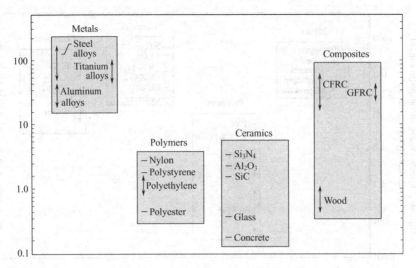

Figure 1.7 Bar-chart of room-temperature resistance to fracture (i.e., fracture toughness) for various metals, ceramics, polymers, and composite materials

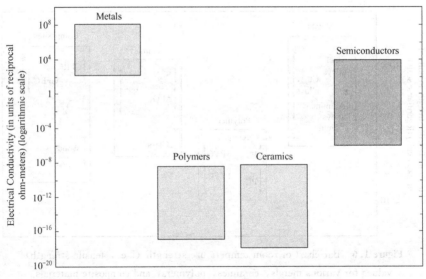

Figure 1.8 Bar-chart of room temperature electrical conductivity ranges for metals, ceramics, polymers, and semiconducting materials

ity (Figure 1.8) and heat, and are not transparent to visible light; a polished metal surface has a lustrous appearance. In addition, some of the metals (viz., Fe, Co, and Ni) have desirable magnetic properties. Furthermore, the types and applications of metals and their alloys are discussed in Chapter 4.

Ceramics

Ceramics are compounds between metallic and nonmetallic elements; they are most frequently oxides, nitrides, and carbides. For example, some of the common ceramic materials include aluminum oxide (or alumina, Al_2O_3), silicon dioxide (or silica, SiO_2), silicon carbide (SiC), silicon nitride (Si_3N_4), and, in addition, what some refer to as the traditional ceramics—those composed of clay minerals (i.e., porcelain), as well as cement, and glass. With regard to mechanical behavior, ceramic materials are relatively stiff and strong—stiffnesses and strengths are comparable to those of the metals (Figure 1.5 and Figure 1.6). In addition, ceramics are typically very hard. On the other hand, they are extremely brittle (lack ductility), and are highly susceptible to fracture (Figure 1.7). These materials are typically insulative to the passage of heat and electricity (i.e., have low electrical conductivities, Figure 1.8), and are more resistant to high temperatures and harsh environments than metals and polymers. With regard to optical characteristics, ceramics may be transparent, translucent, or opaque, and some of the oxide ceramics (e.g., Fe_3O_4) exhibit magnetic behavior. The characteristics, types, and applications of this class of materials are discussed in Chapter 5.

Composites

A composite is composed of two (or more) individual materials, which come from the categories discussed above—viz., metals, ceramics, and polymers. The design goal of a composite is to achieve a combination of properties that is not displayed by any single material, and also to incorporate the best characteristics of each of the component materials. A large number of composite types exist that are represented by different combinations of metals, ceramics, and polymers. Furthermore, some naturally-occurring materials are also considered to be composites—for example, wood and bone. However, most of those we consider in our discussions are synthetic (or man-made) composites.

One of the most common and familiar composites is fiberglass, in which small glass fibers are embedded within a polymeric material (normally an epoxy or polyester). The glass fibers are relatively strong and stiff (but also brittle), whereas the polymer is ductile (but also weak and flexible). Thus, the resulting fiberglass is relatively stiff, strong, (Figure 1.5 and Figure 1.6) flexible, and ductile. In addition, it has a low density (Figure 1.4).

Another of these technologically important materials is the "carbon fiber-reinforced polymer" (or "CFRP") composite—carbon fibers that are embedded within a polymer. These materials are stiffer and stronger than the glass fiber-reinforced materials (Figure 1.5 and Figure 1.6), yet they are more expensive. The CFRP composites are used in some aircraft and aerospace applications, as well as high-tech sporting equipment (e.g., bicycles, golf clubs, tennis rackets, and skis/snowboards). Chapter 6 is devoted to a discussion of these interesting materials.

1.5 Advanced Materials

Materials that are utilized in high-technology (or high-tech) applications are sometimes termed advanced materials. By high technology we mean a device or product that operates or functions using relatively intricate and sophisticated principles; examples include electronic equipment (camcorders, CD/DVD players, etc.), computers, fiber-optic systems, spacecraft, aircraft, and military rocketry. These advanced materials are typically traditional materials whose properties have been enhanced, and, also newly developed, high-performance materials. Furthermore, they may be of all material types (e.g., metals, ceramics, polymers), and are normally expensive. Advanced materials include semiconductors, biomaterials, and what we may term "materials of the future" (that is, smart materials and nanoengineered materials), which we discuss below. The properties and applications of a number of these advanced materials—for example, materials that are used for lasers, integrated circuits, magnetic information storage, liquid crystal displays (LCDs), and fiber optics—are also discussed in subsequent chapters.

Semiconductors

Semiconductors have electrical properties that are intermediate between the electrical conductors (viz. metals and metal alloys) and insulators (viz. ceramics and polymers)—Figure 1.8, discussed in Chapter 8. Furthermore, the electrical characteristics of these materials are extremely sensitive to the presence of minute concentrations of impurity atoms, for which the concentrations may be controlled over very small spatial regions. Semiconductors have made possible the advent of integrated circuitry that has totally revolutionized the electronics and computer industries (not to mention our lives) over the past three decades.

Biomaterials

Biomaterials are employed in components implanted into the human body for replacement of diseased or damaged body parts. These materials must not produce toxic substances and must be compatible with body tissues (i.e., must not cause adverse biological reactions). All of the above materials—metals, ceramics, polymers, composites, and semiconductors—may be used as biomaterials. For example, some of the biomaterials that are utilized in artificial hip replacements are discussed in Chapter 9.

1.6 Modern Materials' Needs

In spite of the tremendous progress that has been made in the discipline of materials science and engineering within the past few years, there still remain technological challenges, including the development of even more sophisticated and specialized materials, as well as consideration of the environmental impact of materials production. Some comment is appropriate relative to these issues so as to round out this perspective.

Nuclear energy holds some promise, but the solutions to the many problems that remain will necessarily involve materials, from fuels to containment structures to facilities for the disposal of radioactive waste.

Significant quantities of energy are involved in transportation. Reducing the weight of transportation vehicles (automobiles, aircraft, trains, etc.), as well as increasing engine operating temperatures, will enhance fuel efficiency. New high-strength, low-density structural materials remain to be developed, as well as materials that have higher-temperature capabilities, for use in engine components.

Furthermore, there is a recognized need to find new, economical sources of energy and to use present resources more efficiently. Materials will undoubtedly play a significant role in these developments. For example, the direct conversion of solar into electrical energy has been demonstrated. Solar cells employ some rather complex and expensive materials. To ensure a viable technology, materials that are highly efficient in this conversion process yet less costly must be developed.

The hydrogen fuel cell is another very attractive and feasible energy-conversion technology that has the advantage of being non-polluting. It is just beginning to be implemented in batteries for electronic devices, and holds promise as the power plant for automobiles. New materials still need to be developed for more efficient fuel cells, and also for better catalysts to be used in the production of hydrogen.

Furthermore, environmental quality depends on our ability to control air and water pollution. Pollution control techniques employ various materials. In addition, materials processing and refinement methods need to be improved so that they produce less environmental degradation—that is, less pollution and less despoil age of the landscape from the mining of raw materials. Also, in some materials manufacturing processes, toxic substances are produced, and the ecological impact of their disposal must be considered.

Many materials that we use are derived from resources that are nonrenewable —that is, not capable of being regenerated. These include polymers, for which the prime raw material is oil, and some metals. These nonrenewable resources are gradually becoming depleted, which necessitates: ①the discovery of additional reserves, ②the development of new materials having comparable properties with less adverse environmental impact, and/or ③increased recycling efforts and the development of new recycling technologies. As a consequence of the economics of not only production but also environmental impact and ecological factors, it is becoming increasingly important to consider the "cradle-to-grave" life cycle of materials relative to the overall manufacturing process.

References

[1] Ashby, M. F. and D. R. H. Jones, *Engineering Materials 1*, *An Introduction to Their Properties and Applications*, 3rd edition, Butterworth Heinemann, Woburn, UK, 2005.

[2] Ashby, M. F. and D. R. H. Jones, *Engineering Materials 2*, *An Introduction to Microstructures*, *Processing and Design*, 3rd edition, Butterworth Heinemann, Woburn, UK, 2005.

[3] Askeland, D. R. and P. P. Phulé, *The Science and Engineering of Materials*, 5th edition, Nelson (a division of Thomson Canada), Toronto, 2006.

[4] Baillie, C. and L. Vanasupa, *Navigating the Materials World*, Academic Press, San Diego, CA, 2003.

[5] Flinn, R. A. and P. K. Trojan, *Engineering Materials and Their Applications*, 4th edition, John Wiley & Sons, New York, 1994.

[6] Jacobs, J. A. and T. F. Kilduff, *Engineering Materials Technology*, 5th edition, Prentice Hall PTR, Paramus, NJ, 2005.

[7] Mangonon, P. L., *The Principles of Materials Selection for Engineering Design*, Prentice Hall PTR, Paramus, NJ, 1999.

[8] McMahon, C. J., Jr., *Structural Materials*, Merion Books, Philadelphia, 2004.

[9] Murray, G. T., *Introduction to Engineering Materials—Behavior, Properties, and Selection*, Mar-

cel Dekker, Inc., New York, 1993.

[10] Ralls, K. M., T. H. Courtney, and J. Wulff, *Introduction to Materials Science and Engineering*, John Wiley & Sons, New York, 1976.

[11] Schaffer, J. P., A. Saxena, S. D. Antolovich, T. H. Sanders, Jr., and S. B. Warner, *The Science and Design of Engineering Materials*, 2nd edition, WCB/McGraw-Hill, New York, 1999.

[12] Shackelford, J. F., *Introduction to Materials Science for Engineers*, 6th edition, Prentice Hall PTR, Paramus, NJ, 2005.

[13] Smith, W. F. and J. Hashemi, *Principles of Materials Science and Engineering*, 4th edition, McGraw-Hill Book Company, New York, 2006.

[14] Van Vlack, L. H., *Elements of Materials Science and Engineering*, 6th edition, Addison-Wesley Longman, Boston, MA, 1989.

[15] White, M. A., *Properties of Materials*, Oxford University Press, New York, 1999.

Chapter 2   The Structure of Crystalline Solids

Learning Objectives
1. (a) *Briefly describe ionic, covalent, metallic, hydrogen, and van der Waals bonds.*
   (b) *Note which materials exhibit each of these bonding types.*
2. *Draw unit cells for face-centered cubic, body-centered cubic, and hexagonal close-packed crystal structures.*
3. *Derive the relationships between unit cell edge length and atomic radius for face-centered cubic and body-centered cubic crystal structures.*
4. *Given three direction index integers, sketch the direction corresponding to these indices within a unit cell.*
5. *Specify the Miller indices for a plane that has been drawn within a unit cell.*
6. *Describe how face-centered cubic and hexagonal close-packed crystal structures may be generated by the stacking of close-packed planes of atoms.*
7. *Distinguish between single crystals and polycrystalline materials.*
8. *Define isotropy and anisotropy with respect to material properties.*
9. *Describe both vacancy and self-interstitial crystalline defects.*
10. *Calculate the equilibrium number of vacancies in a material at some specified temperature, given the relevant constants.*
11. *Name the two types of solid solutions, and provide a brief written definition and/or schematic sketch of each.*
12. *Given the masses and atomic weights of two or more elements in a metal alloy, calculate the weight percent and atom percent for each element.*
13. *For each of edge, screw, and mixed dislocations: (a) describe and make a drawing of the dislocation. (b) note the location of the dislocation line, and (c) indicate the direction along which the dislocation line extends.*
14. *Describe the atomic structure within the vicinity of (a) a grain boundary, and (b) a twin boundary.*

2.1   Atomic Structure and Interatomic Bonding

Some of the important properties of solid materials depend on geometrical atomic arrangements, and also the interactions that exist among constituent atoms or molecules. This chapter, by way of preparation for subsequent discussions, considers several fundamental and important concepts, namely: atomic structure, electron configurations in atoms and the periodic table, and the various types of primary and secondary interatomic bonds that hold together the atoms comprising a solid. These topics are reviewed briefly, under the assumption that some of the material is familiar to the reader.

2.1.1 Fundamental Concepts

Each atom consists of a very small nucleus composed of protons and neutrons which is encircled by moving electrons. Both electrons and protons are electrically charged, the charge magnitude being 1.60×10^{-19} C, which is negative in sign for electrons and positive for protons; neutrons are electrically neutral. Masses for these subatomic particles are infinitesimally small; protons and neutrons have approximately the same mass, 1.67×10^{-27} kg, which is significantly larger than that of an electron, 9.11×10^{-31} kg. Each chemical element is characterized by the number of protons in the nucleus, or the **atomic number** (Z). For an electrically neutral or complete atom, the atomic number also equals the number of electrons. This atomic number ranges in integral units from 1 for hydrogen to 92 for uranium, the highest of the naturally occurring elements.

The *atomic mass* (A) of a specific atom may be expressed as the sum of the masses of protons and neutrons within the nucleus. Although the number of protons is the same for all atoms of a given element, the number of neutrons (N) may be variable. Thus atoms of some elements have two or more different atomic masses, which are called **isotopes**. The **atomic weight** of an element corresponds to the weighted average of the atomic masses of the atom's naturally occurring isotopes. The **atomic** may be used for computations of atomic weight. A scale has been established whereby 1 amu is defined as 1/12 of the atomic mass of the most common isotope of carbon, carbon 12 (^{12}C) ($A=12.00000$). Within this scheme, the masses of protons and neutrons are slightly greater than unity, and

$$A \cong Z + N \tag{2.1}$$

The atomic weight of an element or the molecular weight of a compound may be specified on the basis of amu per atom (molecule) or mass per mole of material. In one **mole** of a substance there are 6.023×10^{23} (Avogadro's number) atoms or molecules. These two atomic weight schemes are related through the following equation:

$$1 \text{amu/atom(or molecule)} = 1 \text{g/mol}$$

For example, the atomic weight of iron is 55.85 amu/atom, or 55.85 g/mol. Sometimes use of amu per atom or molecule is convenient; on other occasions, g (or kg)/mol is preferred; the latter is used in this book.

2.1.2 Bonding Forces and Energies

An understanding of many of the physical properties of materials is predicated on a knowledge of the interatomic forces that bind the atoms together. Perhaps the principles of atomic bonding are best illustrated by considering the interaction between two isolated atoms as they are brought into close proximity from an infinite separation. At large distances, the interactions are negligible, but as the atoms approach, each exerts forces on the other. These forces are of two types, attractive and repulsive, and the magnitude of each is a function of the separation or interatomic distance. The origin of an attractive force F_A depends on the particular type of bonding that exists between the two atoms. The magnitude of the attractive force varies with the distance, as represented schematically in Figure 2.1(a). Ultimately, the outer electron shells of the two atoms begin to overlap, and a strong repulsive force F_R comes into play. The net force F_N between the two atoms is just the sum of both attractive and repulsive components; that is, which is also a function of the interatomic separation, as also plotted in Figure 2.1(a). When F_A and F_R balance, or become equal, there is no net force; that is

$$F_A + F_R = 0 \qquad (2.2)$$

Then a state of equilibrium exists. The centers of the two atoms will remain separated by the equilibrium spacing r_0, as indicated in Figure 2.1(a). For many atoms, r_0 is approximately 0.3 nm. Once in this position, the two atoms will counteract any attempt to separate them by an attractive force, or to push them together by a repulsive action. Sometimes it is more convenient to work with the potential energies between two atoms instead of forces. Mathematically, energy (E) and force (F) are related as

$$E = \int F dr \qquad (2.3)$$

or, for atomic systems

$$E_N = \int_\infty^r F_N dr \qquad (2.4)$$

$$= \int_\infty^r F_A dr + \int_\infty^r F_R dr \qquad (2.5)$$

$$= E_A + E_R \qquad (2.6)$$

in which E_N, E_A, and E_R are respectively the net, attractive, and repulsive energies for two isolated and adjacent atoms. Figure 2.1(b) plots attractive, repulsive, and net potential energies as a function of interatomic separation for two atoms. The net curve, which is again the sum of the other two, has a potential energy trough or well around its minimum. Here, the same equilibrium spacing, r_0, corresponds to the separation distance at the minimum of the potential energy curve. The **bonding energy** for these two atoms, E_0, corresponds to the energy at this minimum point [also shown in Figure 2.1(b)]; it represents the energy that would be required to separate these two atoms to an infinite separation.

Although the preceding treatment has dealt with an ideal situation involving only two atoms, a similar yet more complex condition exists for solid materials because force and energy interactions among many atoms must be considered. Nevertheless, a bonding energy, analogous to E_0 above, may be associated with each atom. The magnitude of this bonding energy and the shape of the energy-versus-interatomic separation curve vary from material to material, and they both depend on the type of atomic bonding. Furthermore, a number of material properties depend on E_0, the curve shape, and bonding type. For example, materials having large bonding energies typically also have high melting temperatures; at room temperature, solid substances are formed for large bonding energies, whereas for small energies the gaseous state is favored; liquids prevail when the energies are of intermediate magnitude. In addition, the mechanical stiffness (or modulus of elasticity) of a material is dependent on the shape of its force-versus-interatomic separation curve. The slope for a relatively stiff material at the r_0 position on the curve will be quite steep; slopes are shallower for more flexible materials. Furthermore, how much a material expands upon heating or contracts upon cooling (that is, its linear coefficient of thermal expansion) is related to the shape of its E_0-versus-r_0 curve. A deep and narrow "trough", which typically occurs for materials having large bonding energies, normally correlates with a low coefficient of thermal expansion and relatively small dimensional alterations for changes in temperature.

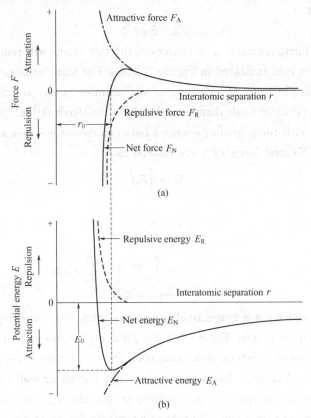

Figure 2.1 (a) The dependence of repulsive, attractive, and net forces on interatomic separation for two isolated atoms. (b) The dependence of repulsive, attractive, and net potential energies on interatomic separation for two isolated atoms. (Adapted from William D. Callister Jr., *Fundamentals of Materials Science and Engineering*, 7th Edition, John Wiley & Sons, Inc., 2007, p. 25.)

2.1.3 Atomic Bonding in Solids
2.1.3.1 Ionic Bonding

Perhaps **ionic bonding** is the easiest to describe and visualize. It is always found in compounds that are composed of both metallic and nonmetallic elements, elements that are situated at the horizontal extremities of the periodic table. Atoms of a metallic element easily give up their valence electrons to the nonmetallic atoms. In the process all the atoms acquire stable or inert gas configurations and, in addition, an electrical charge; that is, they become ions. Sodium chloride (NaCl) is the classical ionic material. A sodium atom can assume the electron structure of neon (and a net single positive charge) by a transfer of its one valence 3s electron to a chlorine atom. After such a transfer, the chlorine ion has a net negative charge and an electron configuration identical to that of argon. In sodium chloride, all the sodium and chlorine exist as ions. This type of bonding is illustrated schematically in Figure 2.2.

The attractive bonding forces are **coulombic**; that is, positive and negative ions, by virtue of their net electrical charge, attract one another.

Ionic bonding is termed nondirectional, that is, the magnitude of the bond is equal in all directions around an ion. It follows that for ionic materials to be stable, all positive ions must have as nearest neighbors negatively charged ions in a three-dimensional scheme, and vice versa. The predominant bonding in ceramic materials is ionic. Some of the ion arrangements for these materials are discussed in following section. Bonding energies, which generally range between 600 and 1500 kJ/mol (3 and 8 eV/atom), are relatively large, as reflected in high melting temperatures. Table 2.1 contains bonding energies and melting temperatures for Various substances.

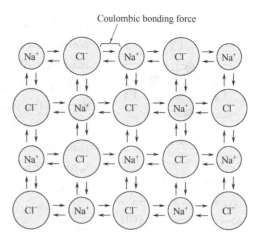

Figure 2.2 Schematic representation of ionic bonding in sodium chloride (NaCl). (Adapted from William D. Callister Jr., *Fundamentals of Materials Science and Engineering*, 7th Edition, John Wiley & Sons, Inc., 2007, p. 27.)

Table 2.1 Bonding Energies and Melting Temperatures for Various Substances

Bonding Type	Substance	Bonding Energy		Melting Temperature /℃
		kJ/mol (kcal/mol)	eV/Atom, Ion, Molecule	
Ionic	NaCl	640(153)	3.3	801
	MgO	1000(239)	5.2	2800
Covalent	Si	450(108)	4.7	1410
	C(diamond)	713(170)	7.4	>3550
Metallic	Hg	68(16)	0.7	−39
	Al	324(77)	3.4	660
	Fe	406(97)	4.2	1538
	W	849(203)	8.8	3410
van der Waals	Ar	7.7(1.8)	0.08	−189
	Cl_2	31(7.4)	0.32	−101
Hydrogen	NH_3	35(8.4)	0.36	−78
	H_2O	51(12.2)	0.52	0

Ionic materials are characteristically hard and brittle and, furthermore, electrically and thermally insulative. As discussed in subsequent chapters, these properties are a direct consequence of electron configurations and/or the nature of the ionic bond.

2.1.3.2 Covalent Bonding

In **covalent bonding** stable electron configurations are assumed by the sharing of electrons between adjacent atoms. Two atoms that are covalently bonded will each contribute at least

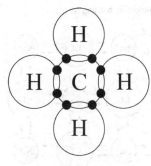

Figure 2.3 Schematic representation of covalent bonding in methane (CH₄)

one electron to the bond, and the shared electrons may be considered to belong to both atoms. Covalent bonding is schematically illustrated in Figure 2.3 for a molecule of methane (CH_4). The carbon atom has four valence electrons, whereas each of the four hydrogen atoms has a single valence electron. Each hydrogen atom can acquire a helium electron configuration (two 1s valence electrons) when the carbon atom shares with it one electron. The carbon now has four additional shared electrons, one from each hydrogen, for a total of eight valence electrons, and the electron structure of neon. The covalent bond is directional; that is, it is between specific atoms and may exist only in the direction between one atom and another that participates in the electron sharing (Figure 2.4).

Many nonmetallic elemental molecules (H_2, Cl_2, F_2, etc.) as well as molecules containing dissimilar atoms, such as CH_4, H_2O, HNO_3, and HF, are covalently bonded. Furthermore, this type of bonding is found in elemental solids such as diamond (carbon), silicon, and germanium and other solid compounds composed of elements that are located on the right-hand side of the periodic table, such as gallium arsenide (GaAs), indium antimonide (InSb), and silicon carbide (SiC).

The number of covalent bonds that is possible for a particular atom is determined by the number of valence electrons. For N' valence electrons, an atom can covalently bond with at most $8-N'$ other atoms. For example, $N'=7$ for chlorine, and $8-N'=1$, which means that one Cl atom can bond to only one other atom, as in Cl_2. Similarly, for carbon, $N'=4$, and each carbon atom has $8-4$, or four, electrons to share. Diamond is simply the three-dimensional interconnecting structure wherein each carbon atom covalently bonds with four other carbon atoms.

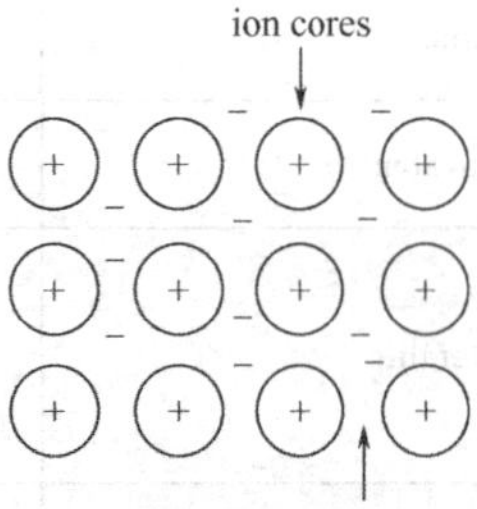

Covalent bonds may be very strong, as in diamond, which is very hard and has a very high melting temperature, >3550℃, or they may be very weak, as with bismuth, which melts at about 270℃. Bonding energies and melting temperatures for a few covalently bonded materials are presented in Table 2.1. Polymeric materials typify this bond, the basic molecular structure being a long chain of carbon atoms that are covalently bonded together with two of their available four bonds per atom. The remaining two bonds normally are shared with other atoms, which also covalently bond. Polymeric molecular structures are discussed in detail in Chapter 3.

Figure 2.4 Schematic representation of metallic bonding

It is possible to have interatomic bonds that are partially ionic and partially covalent, and, in fact, very few compounds exhibit pure ionic or covalent bonding. For a compound, the degree of either bond type depends on the relative positions of the constituent atoms in the periodic table or the difference in their electronegativities. The wider the separation (both horizontally—relative to Group ⅣA—and vertically) from the lower left to the upper-right-hand corner (i. e., the greater the difference in electronegativity), the more ionic the bond. Conversely, the closer the atoms are together (i. e., the smaller the difference in electronegativity), the greater the degree of covalency. The percent ionic character of a bond between elements A and B (A being the most electronegative) may be approximated by the expression

$$\% \text{ ionic character} = \{1 - \exp[-(0.25)(X_A - X_B)^2]\} \times 100 \qquad (2.7)$$

where X_A and X_B are the electronegativities for the respective elements.

2.1.3.3 Metallic Bonding

Metallic bonding, the final primary bonding type, is found in metals and their alloys. A relatively simple model has been proposed that very nearly approximates the bonding scheme. Metallic materials have one, two, or at most, three valence electrons. With this model, these valence electrons are not bound to any particular atom in the solid and are more or less free to drift throughout the entire metal. They may be thought of as belonging to the metal as a whole, or forming a "sea of electrons" or an "electron cloud." The remaining non valence electrons and atomic nuclei form what are called *ion cores*, which possess a net positive charge equal in magnitude to the total valence electron charge per atom. Figure 2.4 is a schematic illustration of metallic bonding. The free electrons shield the positively charged ion cores from mutually repulsive electrostatic forces, which they would otherwise exert upon one another; consequently the metallic bond is nondirectional in character. In addition, these free electrons act as a "glue" to hold the ion cores together. Bonding energies and melting temperatures for several metals are listed in Table 2.1. Bonding may be weak or strong; energies range from 68 kJ/mol (0.7 eV/atom) for mercury to 850 kJ/mol (8.8 eV/atom) for tungsten. Their respective melting temperatures are $-39°C$ and $3410°C$ ($-38°F$ and $6170°F$). Metallic bonding is found for Group ⅠA and ⅡA elements in the periodic table, and, in fact, for all elemental metals.

Some general behaviors of the various material types (i. e., metals, ceramics, polymers) may be explained by bonding type. For example, metals are good conductors of both electricity and heat, as a consequence of their free electrons. By way of contrast, ionically and covalently bonded materials are typically electrical and thermal insulators, due to the absence of large numbers of free electrons. Furthermore, we note that at room temperature, most metals and their alloys fail in a ductile manner; that is, fracture occurs after the materials have experienced significant degrees of permanent deformation. This behavior is explained in terms of deformation mechanism, which is implicitly related to the characteristics of the metallic bond. Conversely, at room temperature ionically bonded materials are intrinsically brittle as a consequence of the electrically charged nature of their component ions.

2.1.3.4 Secondary Bonding or van der Waals Bonding

Secondary, van der Waals, or physical bonds are weak in comparison to the primary or chemical ones; bonding energies are typically on the order of only 10 kJ/mol (0.1eV/atom). Secondary bonding exists between virtually all atoms or molecules, but its presence may be obscured if any of the three primary bonding types is present. Secondary bonding is evidenced for the inert gases, which have stable electron structures, and, in addition, between molecules in molecular structures that are covalently bonded.

Secondary bonding forces arise from atomic or molecular **dipoles**. In essence, an electric dipole exists whenever there is some separation of positive and negative portions of an atom or molecule. The bonding results from the Coulombic attraction between the positive end of one dipole and the negative region of an adjacent one, as indicated in Figure 2.5. Dipole interactions occur between induced dipoles, between induced dipoles and polar molecules (which have permanent dipoles), and between polar molecules. **Hydrogen bonding**, a special type of secondary bonding, is found to exist between some molecules that have hydrogen as one of the constituents. These bonding mechanisms are now discussed briefly.

Figure 2.5 Schematic representation of van der Waals bonding between two dipoles

Fluctuating induced dipole bonds

A dipole may be created or induced in an atom or molecule that is normally electrically symmetric; that is, the overall spatial distribution of the electrons is symmetric with respect to the positively charged nucleus, as shown in Figure 2.6(a). All atoms are experiencing constant vibrational motion that can cause instantaneous and short-lived distortions of this electrical symmetry for some of the atoms or molecules, and the creation of small electric dipoles, as represented in Figure 2.6(b). One of these dipoles can in turn produce a displacement of the electron distribution of an adjacent molecule or atom, which induces the second one also to become a dipole that is then weakly attracted or bonded to the first; this is one type of van der Waals bonding. These attractive forces may exist between large numbers of atoms or molecules, which forces are temporary and fluctuate with time. The liquefaction and, in some cases, the solidification of the inert gases and other electrically neutral and symmetric molecules such as H_2 and Cl_2 are realized because of this type of bonding. Melting and boiling temperatures are extremely low in materials for which induced dipole bonding predominates; of all possible intermolecular bonds, these are the weakest. Bonding energies and melting temperatures for argon and chlorine are also tabulated in Table 2.1.

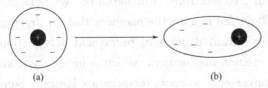

Figure 2.6 Schematic representation of (a) an electrically symmetric atom and (b) an induced atomic dipole

Polar molecule-induced dipole bonds

Permanent dipole moments exist in some molecules by virtue of an asymmetrical arrangement of positively and negatively charged regions; such molecules are termed **polar molecules**. Figure 2.7 is a schematic representation of a polar hydrogen chloride molecule; a permanent dipole moment arises from net positive and negative charges that are respectively associated with the hydrogen and chlorine ends of the HCl molecule. Polar molecules can also induce dipoles in adjacent nonpolar molecules, and a bond will form as a result of attractive forces between the two molecules. Furthermore, the magnitude of this bond will be greater than for fluctuating induced dipoles.

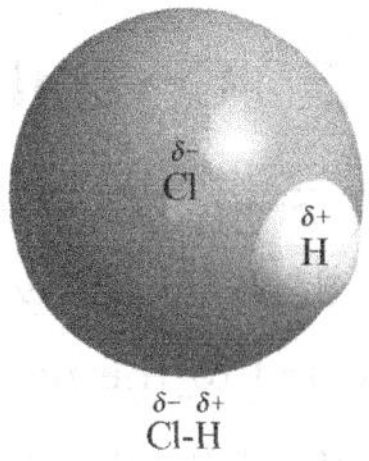

Figure 2.7 Schematic representation of a polar hydrogen chloride molecule (HCl)

Permanent Dipole Bonds

Van der Waals forces will also exist between adjacent polar molecules. The associated bonding energies are significantly greater than for bonds involving induced dipoles. The strongest secondary bonding type, the hydrogen bond, is a special case of polar molecule bonding. It occurs between molecules in which hydrogen is covalently bonded to fluorine (as in HF), oxygen (as in H_2O), and nitrogen (as in NH_3). For each H-F, H-O, or H-N bond, the single hydrogen electron is shared with the other atom. Thus, the hydrogen end of the bond

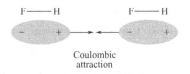

Figure 2.8 Schematic representation of hydrogen bonding in hydrogen fluoride (HF)

is essentially a positively charged bare proton that is unscreened by any electrons. This highly positively charged end of the molecule is capable of a strong attractive force with the negative end of an adjacent molecule, as demonstrated in Figure 2.8 for HF. In essence, this single proton forms a bridge between two negatively charged atoms. The magnitude of the hydrogen bond is generally greater than that of the other types of secondary bonds, and may be as high as 51 kJ/mol (0.52 eV/molecule), as shown in Table 2.1. Melting and boiling temperatures for hydrogen fluoride and water are abnormally high in light of their low molecular weights, as a consequence of hydrogen bonding.

2.1.3.5 Molecules

At the conclusion of this chapter, let us take a moment to discuss the concept of a **molecule** in terms of solid materials. A molecule may be defined as a group of atoms that are bonded together by strong primary bonds. Within this context, the entirety of ionic and metallically bonded solid specimens may be considered as a single molecule. However, this is not the case for many substances in which covalent bonding predominates; these include elemental diatomic molecules (F_2, O_2, H_2, etc.) as well as a host of compounds (H_2O, CO_2, HNO_3, C_6H_6, CH_4, etc.). In the condensed liquid and solid states, bonds between molecules are weak secondary ones. Consequently, molecular materials have relatively low melting and boiling temperatures. Most of those that have small molecules composed of a few atoms are gases at ordinary, or ambient, temperatures and pressures. On the other hand, many of the modern polymers, being molecular materials composed of extremely large molecules, exist as solids; some of their properties are strongly dependent on the presence of van der Waals and hydrogen secondary bonds.

2.2 Crystal Structures

Section 2.1 was concerned primarily with the various types of atomic bonding, which are determined by the electron structure of the individual atoms. The present discussion is devoted to the next level of the structure of materials, specifically, to some of the arrangements that may be assumed by atoms in the solid state. Within this framework, concepts of crystallinity and noncrystallinity are introduced. For crystalline solids the notion of crystal structure is presented, specified in terms of a unit cell. Crystal structures found in both metals and ceramics are then detailed, along with the scheme by which crystallographic directions and planes are expressed. Single crystals, polycrystalline, and noncrystalline materials are considered.

2.2.1 Fundamental Concepts

Solid materials may be classified according to the regularity with which atoms or ions are arranged with respect to one another. A **crystalline** material is one in which the atoms are situated in a repeating or periodic array over large atomic distances; that is, long-range order exists, such that upon solidification, the atoms will position themselves in a repetitive three-dimensional pattern, in which each atom is bonded to its nearest-neighbor atoms. All metals, many ceramic materials, and certain polymers form crystalline structures under normal solidification conditions. For those that do not crystallize, this long-range atomic order is absent; these *noncrystalline* or *amorphous* materials are discussed briefly at the end of this chapter.

Some of the properties of crystalline solids depend on the **crystal structure** of the material, the manner in which atoms, ions, or molecules are spatially arranged. There is an extremely large number of different crystal structures all having long range atomic order; these vary from relatively simple structures for metals, to exceedingly complex ones, as displayed by some of the ceramic and polymeric materials. The present discussion deals with several common metallic and ceramic crystal structures. The next chapter is devoted to structures for polymers.

When describing crystalline structures, atoms (or ions) are thought of as being solid spheres having well-defined diameters. This is termed the *atomic hard sphere model* in which spheres representing nearest-neighbor atoms touch one another. An example of the hard sphere model for the atomic arrangement found in some of the common elemental metals is displayed in Figure 2.9(c). In this particular case all the atoms are identical. Sometimes the term **lattice** is used in the context of crystal structures; in this sense "lattice" means a three-dimensional array of points coinciding with atom positions (or sphere centers).

Unit Cells

The atomic order in crystalline solids indicates that small groups of atoms form a repetitive pattern. Thus, in describing crystal structures, it is often convenient to subdivide the structure into small repeat entities called **unit cells**. Unit cells for most crystal structures are

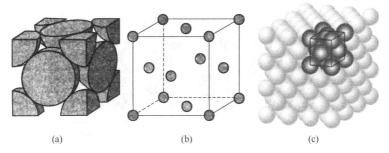

Figure 2.9 For the face centered cubic crystal structure, (a) a hard sphere unit cell representation, (b) a reduced-sphere unit cell, and (c) an aggregate of many atoms

parallelepipeds or prisms having three sets of parallel faces; one is drawn within the aggregate of spheres [Figure 2.9(c)], which in this case happens to be a cube. A unit cell is chosen to represent the symmetry of the crystal structure, wherein all the atom positions in the crystal may be generated by translations of the unit cell integral distances along each of its edges. Thus, the unit cell is the basic structural unit or building block of the crystal structure and defines the crystal structure by virtue of its geometry and the atom positions within. Convenience usually dictates that parallelepiped corners coincide with centers of the hard sphere atoms. Furthermore, more than a single unit cell may be chosen for a particular crystal structure; however, we generally use the unit cell having the highest level of geometrical symmetry.

2.2.2 Metallic Crystal Structures and Crystal Systems

The atomic bonding in this group of materials is metallic, and thus nondirectional in nature. Consequently, there are no restrictions as to the number and position of nearest-neighbor atoms; this leads to relatively large numbers of nearest neighbors and dense atomic packings for most metallic crystal structures. Also, for metals, using the hard sphere model for the crystal structure, each sphere represents an ion core. Table 2.2 presents the atomic radii for a number of metals. Three relatively simple crystal structures are found for most of the common metals: simple cubic, body-centered cubic, face-centered cubic, and hexagonal close-packed.

2.2.2.1 The Simple Cubic Crystal Structure

The crystal structure found for many metals has a unit cell of cubic geometry. The simplest one is the **simple cubic (SC)** crystal structure, with atoms located at each of the corners and the centers of all the cube faces. Figure 2.10(a) shows a hard sphere model for the SC structure. The unit cell demonstrated in Figure 2.10(b) completely describes the structure of the solid, which can be regarded as an almost endless repetition of the unit cell.

The volume of the unit cell is readily calculated from its shape and dimensions. This calculation is particularly easy for a unit cell that is cubic. In this example, the atoms are in contact with each other along the edges of the unit cell. Thus the side of the unit cell has a length of $2r$, where r is the radius of an atom.

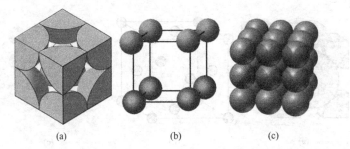

Figure 2.10 For the simple cubic crystal structure, (a) a hard sphere unit cell representation, (b) a reduced-sphere unit cell, and (c) an aggregate of many atoms

Atoms, of course, do not have well-defined bounds, and the radius of an atom is somewhat ambiguous. In the context of crystal structures, the diameter ($2r$) of an atom can be defined as the center-to-center distance between two atoms packed as tightly together as possible. This provides an effective radius for the atom and is sometime called the atomic radius.

Two other important characteristics of a crystal structure are the **coordination number** and the **atomic packing factor (APF)**. For metals, each atom has the same number of nearest-neighbor or touching atoms, which is the coordination number. For SC structure, the coordination number is 6.

The APF is the fraction of solid sphere volume in a unit cell, assuming the atomic hard sphere model, or

$$\text{APF} = \frac{\text{Total sphere volume}}{\text{Total unit cell volume}} = \frac{V_S}{V_C} \tag{2.8}$$

A more challenging task is to determine the number of atoms that included in the unit cell. As described above, an atom is centered on each corner. In this case, however, none of these atoms lies completely within the cell. Part of each atom lies within the unit cell and the remainder lies outside the unit cell. In determining the number of atoms inside the unit cell, one must count only that portion of an atom that actually lies within the unit cell.

For SC structure, there is only one atom included in one unit cell, hence the atomic packing factor is 0.52.

2.2.2.2 The Body-Centered Cubic Crystal Structure

Another common metallic crystal structure also has a cubic unit cell with atoms located at all eight corners and a single atom at the cube center. This is called a **body-centered cubic (BCC)** crystal structure. A collection of spheres depicting this crystal structure is shown in Figure 2.10(c), whereas Figure 2.10(a) and Figure 2.10(b) are diagrams of BCC unit cells with the atoms represented by hard sphere and reduced-sphere models, respectively. Center and corner atoms touch one another along cube diagonals, and unit cell length a and atomic radius R are related through

$$a = 4R/\sqrt{3} \tag{2.9}$$

Chromium, iron, tungsten, as well as several other metals listed in Table 2.2 exhibit a

BCC structure.

Table 2.2　Atomic Radii and Crystal Structures for 16 Metals

Metal	Crystal Structure	Atomic Radius/nm	Metal	Crystal Structure	Atomic Radius/nm
Aluminum(Al)	FCC	0.1431	Molybdenum(Mo)	BCC	0.1363
Cadmium(Cd)	HCP	0.1490	Nickel(Ni)	FCC	0.1246
Chromium(Cr)	BCC	0.1249	Platinum(Pt)	FCC	0.1387
Cobalt(Co)	HCP	0.1253	Silver(Ag)	FCC	0.1445
Copper(Cu)	FCC	0.1278	Tantalum(Ta)	BCC	0.1430
Gold(Au)	FCC	0.1442	Titanium(Ti,a)	HCP	0.1445
Iron(a)	BCC	0.1241	Tungsten(W)	BCC	0.1371
Lead(Pd)	FCC	0.1750	Zinc(Zn)	HCP	0.1332

Two atoms are associated with each BCC unit cell: the equivalent of one atom from the eight corners, each of which is shared among eight unit cells, and the single center atom, which is wholly contained within its cell. In addition, corner and center atom positions are equivalent. The coordination number for the BCC crystal structure is 8; each center atom has as nearest neighbors its eight corner atoms (Figure 2.11).

For the BCC structure, the atomic packing factor is 0.68.

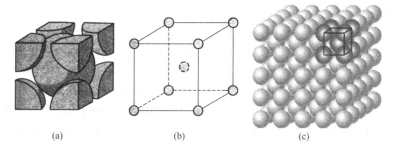

Figure 2.11　For the body-centered cubic crystal structure, (a) a hard sphere unit cell representation, (b) a reduced-sphere unit cell, and (c) an aggregate of many atoms. [Figure (c) from W. G. Moffatt, G. W. Pearsall, and J. Wulff, *The Structure and Properties of Materials*, Vol. I, *Structure*, p. 51.]

2.2.2.3　The Face-Centered Cubic Crystal Structure

The structure shown in Figure 2.9 is aptly called the **face-centered cubic (FCC)** crystal structure. Some of the familiar metals having this crystal structure are copper, aluminum, silver, and gold (see also Table 2.2). Figure 2.9(a) shows a hard sphere model for the FCC unit cell, whereas in Figure 2.9(b) the atom centers are represented by small circles to provide a better perspective of atom positions. The aggregate of atoms in Figure 2.9(c) represents a section of crystal consisting of many FCC unit cells. These spheres or ion cores touch one another across a face diagonal; the cube edge length a and the atomic radius R are related through

$$a = 2R\sqrt{2} \tag{2.10}$$

This result is obtained as an example problem.

For the FCC crystal structure, each corner atom is shared among eight unit cells, whereas a face-centered atom belongs to only two. Therefore, one eighth of each of the eight corner atoms and one half of each of the six face atoms, or a total of four whole atoms, may be assigned to a given unit cell. This is depicted in Figure 2.9(a), where only sphere portions are represented within the confines of the cube. The cell comprises the volume of the cube, which is generated from the centers of the corner atoms as shown in the figure.

Corner and face positions are really equivalent; that is, translation of the cube corner from an original corner atom to the center of a face atom will not alter the cell structure.

For face-centered cubics, the coordination number is 12. This may be confirmed by examination of Figure 2.9(a); the front face atom has four corner nearest-neighbor atoms surrounding it, four face atoms that are in contact from behind, and four other equivalent face atoms residing in the next unit cell to the front, which is not shown.

For the FCC structure, the atomic packing factor is 0.74, which is the maximum packing possible for spheres all having the same diameter. Computation of this APF is also included as an example problem. Metals typically have relatively large atomic packing factors to maximize the shielding provided by the free electron cloud.

Since the coordination number is larger for FCC than BCC, so also is the atomic packing factor for FCC lower—0.74 versus 0.68.

2.2.2.4 The Hexagonal Close-Packed Crystal Structure

Not all metals have unit cells with cubic symmetry; the final common metallic crystal structure to be discussed has a unit cell that is hexagonal. Figure 2.12(a) shows a hard sphere model for this structure, which is termed **hexagonal close-packed (HCP)**; a reduced-sphere unit cell of HCP unit cell is presented in Figure 2.12(b). The top and bottom faces of the unit cell consist of six atoms that form regular hexagons and surround a single atom in the center. Another plane that provides three additional atoms to the unit cell is situated between the top and bottom planes. The atoms in this midplane have as nearest neighbors atoms in both of the adjacent two planes. The equivalent of six atoms is contained in each unit cell; one-sixth of each of the 12 top and bottom face corner atoms, one-half of each of the 2 center face atoms, and all the 3 midplane interior atoms. If a and c represent, respectively, the short and long unit cell dimensions of Figure 2.12(b), the c/a ratio should be 1.633; however, for some HCP metals this ratio deviates from the ideal value.

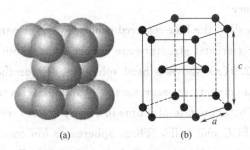

Figure 2.12 For the hexagonal close-packed crystal structure, (a) a hard sphere model, and (b) a reduced-sphere unit cell (a and c represent the short and long edge lengths, respectively)

The coordination number and the atomic packing factor for the HCP crystal structure are the same as for FCC: 12 and 0.74, respectively. The HCP metals include cadmium, magnesium, titanium, and zinc; some of these are listed in Table 2.2.

Example Problem 2.1

Calculate the volume of an FCC unit cell in terms of the atomic radius R.

Solution

In the FCC unit cell illustrated, the atoms touch one another across a face-diagonal the length of which is $4R$.

Since the unit cell is a cube, its volume is a^3, where a is the cell edge length.

From the right triangle on the face

$$a^2 + a^2 = (4R)^2$$

or, solving for a

$$a = 2R\sqrt{2}$$

The FCC unit cell volume V_C may be computed from

$$V_C = a^3 = (2R\sqrt{2})^3 = 16R^3\sqrt{2} \tag{2.11}$$

Example Problem 2.2

Show that the atomic packing factor for the FCC crystal structure is 0.74.

Solution

The APF is defined as the fraction of solid sphere volume in a unit cell, or

$$\text{APF} = \frac{\text{Total sphere volume}}{\text{Total unit cell volume}} = \frac{V_S}{V_C}$$

Both the total sphere and unit cell volumes may be calculated in terms of the atomic radius R. The volume for a sphere is $\frac{4}{3}\pi R^3$, and since there are four atoms per FCC unit cell, the total FCC sphere volume is

$$V_S = (4)\frac{4}{3}\pi R^3 = \frac{16}{3}\pi R^3$$

From Example Problem 2.1, the total unit cell volume is

$$V_C = 16R^3\sqrt{2}$$

Therefore, the atomic packing factor is

$$\text{APF} = \frac{V_S}{V_C} = \frac{\left(\frac{16}{3}\right)\pi R^3}{16R^3\sqrt{2}} = 0.74$$

2.2.2.5 Density Computations—Metals

A knowledge of the crystal structure of a metallic solid permits computation of its theoretical density ρ through the relationship

$$\rho = \frac{nA}{V_C N_A} \tag{2.12}$$

where

n = number of atoms associated with each unit cell

A = atomic weight

V_C = volume of the unit cell

N_A = Avogadro's number (6.023×10²³ atoms/mol)

Example Problem 2.3

Copper has an atomic radius of 0.128 nm (1.28Å), an FCC crystal structure, and an atomic weight of 63.5 g/mol. Compute its theoretical density and compare the answer with its measured density.

Solution

Equation 2.12 is employed in the solution of this problem. Since the crystal structure is FCC, n, the number of atoms per unit cell, is 4. Furthermore, the atomic weight A_{Cu} is given as 63.5g/mol. The unit cell volume V_C for FCC was determined in Example Problem 2.1 as $16R^3\sqrt{2}$, where R, the atomic radius, is 0.128 nm.

Substitution for the various parameters into Equation 2.12 yields

$$\rho = \frac{nA_{Cu}}{V_C N_A} = \frac{nA_{Cu}}{(16R^3\sqrt{2})N_A}$$

$$= \frac{(4 \text{atoms/unit cell})(63.5 \text{ g/mol})}{[16\sqrt{2}(1.28\times10^{-8}\text{cm})^3/\text{unit cell}](6.023\times10^{23})\text{atoms/mol}} = 8.89\text{g/cm}^3$$

The literature value for the density of copper is 8.94 g/cm³, which is in very close agreement with the foregoing result.

2.2.2.6 Crystal Systems

Since there are many different possible crystal structures, it is sometimes convenient to divide them into groups according to unit cell configurations and/or atomic arrangements. One such scheme is based on the unit cell geometry, that is, the shape of the appropriate unit cell parallelepiped without regard to the atomic positions in the cell. Within this framework, an x, y, z coordinate system is established with its origin at one of the unit cell corners; each of the x, y, and z axes coincides with one of the three parallelepiped edges that extend from this corner, as illustrated in Figure 2.13. The unit cell geometry is completely defined in terms of six parameters: the three edge lengths a, b, and c, and the three interaxial angles α, β, and γ. These are indicated in Figure 2.13, and are sometimes termed the **lattice parameters** of a crystal structure.

On this basis there are found crystals having seven different possible combinations of a, b, and c, and α, β, and γ, each of which represents a distinct **crystal system.** These seven crystal systems are cubic, tetragonal, hexagonal, orthorhombic, rhombohedral, monoclinic, and triclinic. The lattice parameter relationships and unit cell sketches for each are represented in Table 2.3. The cubic system, for which $a=b=c$ and $\alpha=\beta=\gamma=90°$, has the greatest degree of symmetry. Least symmetry is displayed

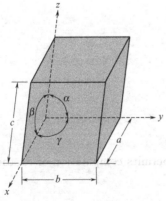

Figure 2.13 A unit cell with x, y, and z coordinate axes, showing axial lengths (a, b, and c) and interaxial angles (α, β, and γ)

by the triclinic system, since $a\neq b\neq c$ and $\alpha\neq\beta\neq\gamma\neq 90°$.

Table 2.3 Lattice Parameter Relationships and Figures Showing Unit Cell Geometries for the Seven Crystal Systems. (Adapted from William D. Callister Jr., *Fundamentals of Materials Science and Engineering*, 7th Edition, John Wiley & Sons, Inc., 2007, p. 47.)

Crystal System	Axial Relationships	Interaxial Angles	Unit Cell Geometry
Cubic	$a=b=c$	$\alpha=\beta=\gamma=90°$	
Hexagonal	$a=b\neq c$	$\alpha=\beta=90°, \gamma=120°$	
Tetragonal	$a=b\neq c$	$\alpha=\beta=\gamma=90°$	
Rhombohedral	$a=b=c$	$\alpha=\beta=\gamma\neq 90°$	
Orthorhombic	$a\neq b\neq c$	$\alpha=\beta=\gamma=90°$	
Monoclinic	$a\neq b\neq c$	$\alpha=\gamma=90°\neq\beta$	
Triclinic	$a\neq b\neq c$	$\alpha\neq\beta\neq\gamma\neq 90°$	

From the discussion of metallic crystal structures, it should be apparent that both FCC and BCC structures belong to the cubic crystal system, whereas HCP falls within hexagonal. The conventional hexagonal unit cell really consists of three parallelepipeds situated as shown in Table 2.3.

2.2.3 Crystallographic Points, Directions, and Planes

When dealing with crystalline materials, it often becomes necessary to specify some particular crystallographic plane of atoms or a crystallographic direction. Labeling conventions have been established in which three integers or indices are used to designate directions and planes. The basis for determining index values is the unit cell, with a coordinate system consisting of three (x, y, and z) axes situated at one of the corners and coinciding with the unit cell edges, as shown in Figure 2.13. For some crystal systems—namely, hexagonal, rhombohedral, monoclinic, and triclinic—the three axes are *not* mutually perpendicular, as in the familiar Cartesian coordinate scheme.

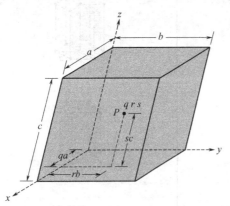

Figure 2.14 The manner in which the q, r, and s coordinates at point P within the unit cell are determined. The q coordinate (which is a fraction) corresponds to the distance qa along the x axis, where a is the unit cell edge length. The respective r and s coordinates for the y and z axes are determined similarly

2.2.3.1 Point Coordinates

The position of any point located within a unit cell may be specified in terms of its coordinates as fractional multiples of the unit cell edge lengths (i.e., in terms of a, b, and c). To illustrate, consider the unit cell and the point P situated therein as shown in Figure 2.14. We specify the position of P in terms of the generalized coordinates q, r, and s where q is some fractional length of a along the x axis, r is some fractional length of b along the y axis, and similarly for s. Thus, the position of P is designated using coordinates $q\,r\,s$ with values that are less than or equal to unity. Furthermore, we have chosen not to separate these coordinates by commas or any other punctuation marks (which is the normal convention).

2.2.3.2 Crystallographic Directions

A crystallographic direction is defined as a line between two points, or a vector. The following steps are utilized in the determination of the three directional indices:

1. A vector of convenient length is positioned such that it passes through the origin of the coordinate system. Any vector may be translated throughout the crystal lattice without alteration, if parallelism is maintained.

2. The length of the vector projection on each of the three axes is determined; *these are measured in terms of the unit cell dimensions a, b, and c.*

3. These three numbers are multiplied or divided by a common factor to reduce them to

the smallest integer values.

4. The three indices, not separated by commas, are enclosed in square brackets, thus: [uvw]. The u, v, and w integers correspond to the reduced projections along the x, y, and z axes, respectively.

For each of the three axes, there will exist both positive and negative coordinates. Thus negative indices are also possible, which are represented by a bar over the appropriate index. For example, the [1 $\bar{1}$1] direction would have a component in the -y direction. Also, changing the signs of all indices produces an antiparallel direction; that is, [$\bar{1}$1 $\bar{1}$] is directly opposite to [1 $\bar{1}$1] . If more than one direction or plane is to be specified for a particular crystal structure, it is imperative for the maintaining of consistency that a positive-negative convention, once established, not be changed. The [100], [110], and [111] directions are common ones; they are drawn in the unit cell shown in Figure 2.15.

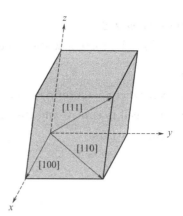

Figure 2.15 The [100], [110], and [111] directions within a unit cell

Example Problem **2.4**

Determine the indices for the direction shown in the accompanying figure.

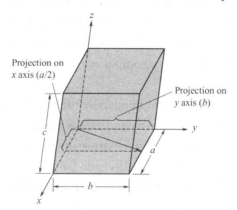

Solution

The vector, as drawn, passes through the origin of the coordinate system, and therefore no translation is necessary. Projections of this vector onto the x, y, and z axes are, respectively, $a/2$, b, and $0c$, which become 1/2, 1, and 0 in terms of the unit cell parameters (i.e., when the a, b, and c are dropped). Reduction of these numbers to the lowest set of integers is accompanied by multiplication of each by the factor 2. This yields the integers 1, 2, and 0, which are then enclosed in brackets as [120].

This procedure may be summarized as follows:

Procedure	x	y	z
Projections	a/2	b	0c
Projections(in terms of a,b and c)	1/2	1	0
Reduction	1	2	0
Enclosure		[120]	

Example Problem 2.5

Draw a [110] direction within a cubic unit cell.

Solution

First construct an appropriate unit cell and coordinate axes system. In the accompanying figure the unit cell is cubic, and the origin of the coordinate system, point O, is located at one of the cube corners.

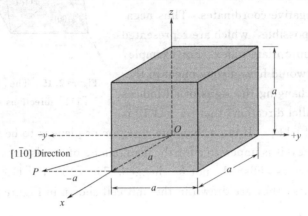

This problem is solved by reversing the procedure of the preceding example. For this [110] direction, the projections along the x, y, z axes are a, $-a$, and $0a$, respectively. This direction is defined by a vector passing from the origin to point P, which is located by first moving along the x axis a units, and from this position, parallel to the y axis $-a$ units, as indicated in the figure. There is no z component to the vector, since the z projection is zero.

For some crystal structures, several nonparallel directions with different indices are actually equivalent; this means that the spacing of atoms along each direction is the same. For example, in cubic crystals, all the directions represented by the following indices are equivalent: [100], [$\bar{1}$00], [010], [0$\bar{1}$0], [001], and [00$\bar{1}$]. As a convenience, equivalent directions are grouped together into a *family*, which are enclosed in angle brackets, thus: <100>. Furthermore, directions in cubic crystals having the same indices without regard to order or sign, for example, [123] and [$\bar{2}$1$\bar{3}$], are equivalent. This is, in general, not true for other crystal systems. For example, for crystals of tetragonal symmetry, [100] and [010] directions are equivalent, whereas [100] and [001] are not.

2.2.3.3 Crystallographic Planes

The orientations of planes for a crystal structure are represented in a similar manner. Again, the unit cell is the basis, with the three-axis coordinate system as represented in Figure 2.13. In all but the hexagonal crystal system, crystallographic planes are specified by three **Miller indices** as (hkl). Any two planes parallel to each other are equivalent and have identical indices. The procedure employed in determination of the h, k, and l index numbers is as follows:

1. If the plane passes through the selected origin, either another parallel plane must be

constructed within the unit cell by an appropriate translation, or a new origin must be established at the corner of another unit cell.

2. At this point the crystallographic plane either intersects or parallels each of the three axes; the length of the planar intercept for each axis is determined in terms of the lattice parameters a, b, and c.

3. The reciprocals of these numbers are taken. A plane that parallels an axis may be considered to have an infinite intercept, and, therefore, a zero index.

4. If necessary, these three numbers are changed to the set of smallest integers by multiplication or division by a common factor.

5. Finally, the integer indices, not separated by commas, are enclosed within parentheses, thus: (hkl).

An intercept on the negative side of the origin is indicated by a bar or minus sign positioned over the appropriate index. Furthermore, reversing the directions of all indices specifies another plane parallel to, on the opposite side of and equidistant from, the origin. Several low-index planes are represented in Figure 2.16.

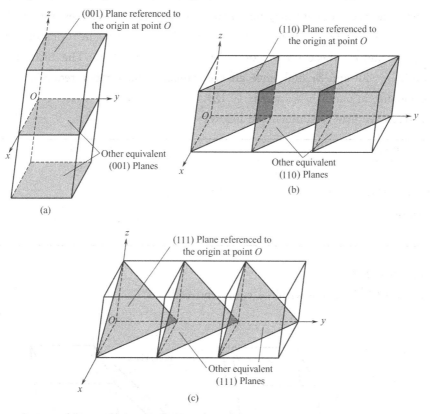

Figure 2.16 Representations of a series each of (a) (001), (b) (110), and (c) (111) crystallographic planes

One interesting and unique characteristic of cubic crystals is that planes and directions having the same indices are perpendicular to one another; however, for other crystal systems there are no simple geometrical relationships between planes and directions having the same indices.

Example Problem 2.6

Determine the Miller indices for the plane shown in the accompanying sketch (*a*).

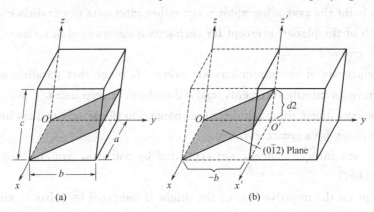

Solution

Since the plane passes through the selected origin O, a new origin must be chosen at the corner of an adjacent unit cell, taken as O' and shown in sketch (*b*). This plane is parallel to the x axis, and the intercept may be taken as ∞a. The y and z axes intersections, referenced to the new origin O', are $-b$ and $c/2$, respectively. Thus, in terms of the lattice parameters a, b, and c, these intersections are ∞, -1, and $1/2$. The reciprocals of these numbers are 0, -1, and 2; and since all are integers, no further reduction is necessary. Finally, enclosure in parentheses yields $(0\,\bar{1}2)$.

These steps are briefly summarized below:

Procedure	x	y	z
Intercepts	$8a$	$-b$	$c/2$
Intercepts(in terms of lattice parameters)	8	-1	$1/2$
Reductions(unnecessary)	0	-1	2
Enclosure		$(0\,\bar{1}2)$	

Example Problem 2.7

Construct a $(0\,\bar{1}1)$ plane within a cubic unit cell.

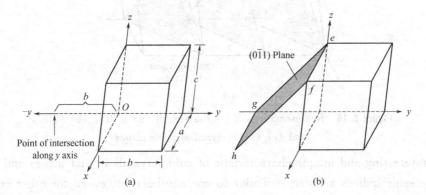

Solution

To solve this problem, carry out the procedure used in the preceding example in reverse order. To begin, the indices are removed from the parentheses, and reciprocals are taken, which yields ∞, −1, and 1. This means that the particular plane parallels the x axis while intersecting the y and z axes at $-b$ and c, respectively, as indicated in the accompanying sketch (a). This plane has been drawn in sketch (b). A plane is indicated by lines representing its intersections with the planes that constitute the faces of the unit cell or their extensions. For example, in this figure, line ef is the intersection between the $(0\bar{1}1)$ plane and the top face of the unit cell; also, line gh represents the intersection between this same $(0\bar{1}1)$ plane and the plane of the bottom unit cell face extended.

Similarly, lines eg and fh are the intersections between $(0\bar{1}1)$ and back and front cell faces, respectively.

2.2.3.4 Atomic Arrangements

The atomic arrangement for a crystallographic plane, which is often of interest, depends on the crystal structure. The (110) atomic planes for FCC and BCC crystal structures are represented in Figure 2.17 and Figure 2.18; reduced-sphere unit cells are also included. Note that the atomic packing is different for each case. The circles represent atoms lying in the crystallographic planes as would be obtained from a slice taken through the centers of the full-sized hard spheres.

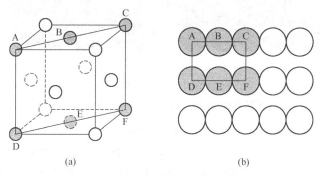

Figure 2.17 (a) Reduced-sphere FCC unit cell with (110) plane. (b) Atomic packing of an FCC (110) plane. Corresponding atom positions from (a) are indicated

A "family" of planes contains all those planes that are crystallographically equivalent—that is, having the same atomic packing; and a family is designated by indices that are enclosed in braces—e.g., {100}. For example, in cubic crystals the (111), $(\bar{1}\bar{1}\bar{1})$, $(\bar{1}11)$, $(1\bar{1}1)$, $(11\bar{1})$, $(\bar{1}1\bar{1})$, $(\bar{1}1\bar{1})$ and $(1\bar{1}\bar{1})$ planes all belong to the {111} family. On the other hand, for tetragonal crystal structures, the {100} family would contain only the (100), $(\bar{1}00)$, (010) and $(0\bar{1}0)$ $(00\bar{1})$ since the (001) and planes are not crystallographically equivalent. Also, in the cubic system only, planes having the same indices, irrespective of order

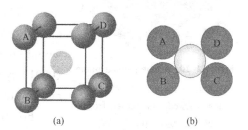

Figure 2.18 (a) Reduced-sphere BCC unit cell with (110) plane. (b) Atomic packing of a BCC (110) plane. Corresponding atom positions from (a) are indicated

and sign, are equivalent. For example, $(1\bar{2}3)$ and $(3\bar{1}2)$ both belong to the $\{123\}$ family.

2.2.3.5 Close-Packed Crystal Structures of Metals

It may be remembered from the discussion on metallic crystal structures that both face-centered cubic and hexagonal close-packed crystal structures have atomic packing factors of 0.74, which is the most efficient packing of equalsized spheres or atoms. In addition to unit cell representations, these two crystal structures may be described in terms of close-packed planes of atoms (i.e., planes having a maximum atom or sphere-packing density); a portion of one such plane is illustrated in Figure 2.19(a). Both crystal structures may be generated by the stacking of these close-packed planes on top of one another; the difference between the two structures lies in the stacking sequence.

Let the centers of all the atoms in one close-packed plane be labeled A. Associated with this plane are two sets of equivalent triangular depressions formed by three adjacent atoms, into which the next close-packed plane of atoms may rest. Those having the triangle vertex pointing up are arbitrarily designated as B positions, while the remaining depressions are those with the down vertices, which are marked C in Figure 2.19(a).

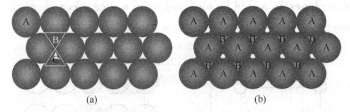

Figure 2.19 (a) A portion of a close-packed plane of atoms; A, B, and C positions are indicated. (b) The ABABAB stacking sequence for close-packed atomic planes

A second close-packed plane may be positioned with the centers of its atoms over either B or C sites; at this point both are equivalent. Suppose that the B positions are arbitrarily chosen; the stacking sequence is termed AB, which is illustrated in Figure 2.19(b). The real distinction between FCC and HCP lies in where the third close-packed layer is positioned. For HCP, the centers of this layer are aligned directly above the original A positions. This stacking sequence, ABABAB..., is repeated over and over. Of course, the ACACAC... arrangement would be equivalent. These close-packed planes for HCP are (0001)-type planes, and the correspondence between this and the unit cell representation is shown in Figure 2.20.

For the face-centered crystal structure, the centers of the third plane are situated over the C sites of the first plane [Figure 2.21(a)] This yields an ABCABCABC... stacking sequence; that is, the atomic alignment repeats every third plane. It is more difficult to correlate the stacking of close-

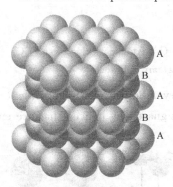

Figure 2.20 Close-packed plane stacking sequence for hexagonal close-packed. (Adapted from W. G. Moffatt, G. W. Pearsall, and J. Wulff, *The Structure and Properties of Materials*, Vol. I, *Structure*, p. 51.)

packed planes to the FCC unit cell. However, this relationship is demonstrated in Figure 2.21(b); these planes are of the (111) type.

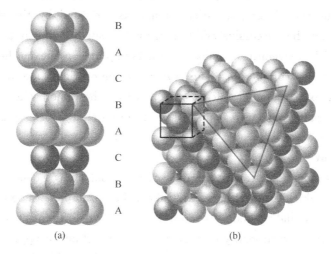

Figure 2.21 (a) Close-packed stacking sequence for face-centered cubic. (b) A corner has been removed to show the relation between the stacking of closepacked planes of atoms and the FCC crystal structure; the heavy triangle outlines a (111) plane. [Figure (b) from W. G. Moffatt, G. W. Pearsall, and J. Wulff, *The Structure and Properties of Materials*, Vol. Ⅰ, *Structure*, p. 51.]

2.2.4 Crystalline and Noncrystalline Materials
2.2.4.1 Single Crystals

For a crystalline solid, when the periodic and repeated arrangement of atoms is perfect or extends throughout the entirety of the specimen without interruption, the result is a **single crystal**. All unit cells interlock in the same way and have the same orientation. Single crystals exist in nature, but they may also be produced artificially. They are ordinarily difficult to grow, because the environment must be carefully controlled.

If the extremities of a single crystal are permitted to grow without any external constraint, the crystal will assume a regular geometric shape having flat faces, as with some of the gem stones; the shape is indicative of the crystal structure. A photograph of several single crystals is shown in Figure 2.22. Within the past few years, single crystals have become extremely important in many of our modern technologies, in particular electronic microcircuits, which employ single crystals of silicon and other semiconductors.

Figure 2.22 Photograph of a rock containing three crystals of pyrite (FeS_2). The crystal structure of pyrite is simple cubic, and this is reflected in the cubic symmetry of its natural crystal facets

2.2.4.2 Polycrystalline Materials

Most crystalline solids are composed of a collection of many small crystals or **grains**;

such materials are termed **polycrystalline**. Various stages in the solidification of a polycrystalline specimen are as follows. Initially, small crystals or nuclei form at various positions. These have random crystallographic orientations, as indicated by the square grids. The small grains grow by the successive addition from the surrounding liquid of atoms to the structure of each. The extremities of adjacent grains impinge on one another as the solidification process approaches completion. The crystallographic orientation varies from grain to grain. Also, there exists some atomic mismatch within the region where two grains meet; this area, called a **grain boundary**, is discussed in more detail in Section 2.3.

2.2.4.3 Anisotropy

The physical properties of single crystals of some substances depend on the crystallographic direction in which measurements are taken. For example, the elastic modulus, the electrical conductivity, and the index of refraction may have different values in the [100] and [111] directions. This directionality of properties is termed **anisotropy**, and it is associated with the variance of atomic or ionic spacing with crystallographic direction. Substances in which measured properties are independent of the direction of measurement are **isotropic.** The extent and magnitude of anisotropic effects in crystalline materials are functions of the symmetry of the crystal structure; the degree of anisotropy increases with decreasing structural symmetry—triclinic structures normally are highly anisotropic. The modulus of elasticity values at [100], [110], and [111] orientations for several materials are presented in Table 2.4.

Table 2.4 Modulus of Elasticity Values for Several Metals at Various Crystallographic Orientations

Metal	Modulus of Elasticity/GPa		
	[100]	[110]	[111]
Aluminum(Al)	63.7	72.6	76.1
Copper(Cu)	66.7	130.3	191.1
Iron(Fe)	125.0	210.5	272.7
Tungsten(W)	384.6	384.6	384.6

For many polycrystalline materials, the crystallographic orientations of the individual grains are totally random. Under these circumstances, even though each grain may be anisotropic, a specimen composed of the grain aggregate behaves isotropically. Also, the magnitude of a measured property represents some average of the directional values. Sometimes the grains in polycrystalline materials have a preferential crystallographic orientation, in which case the material is said to have a "texture."

2.2.4.4 Noncrystalline Solids

It has been mentioned that **noncrystalline** solids lack a systematic and regular arrangement of atoms over relatively large atomic distances. Sometimes such materials are also

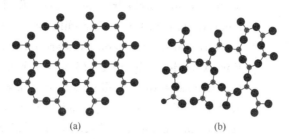

Figure 2.23 Two-dimensional schemes of the structure of (a) crystalline silicon dioxide and (b) noncrystalline silicon dioxide. The blue balls and red balls represent Si and O atoms, respectively

called **amorphous** (meaning literally without form), or supercooled liquids, inasmuch as their atomic structure resembles that of a liquid. An amorphous condition may be illustrated by comparison of the crystalline and noncrystalline structures of the ceramic compound silicon dioxide (SiO_2), which may exist in both states. Figure 2.23(a) and Figure 2.23(b) present two-dimensional schematic diagrams for both structures of SiO_2, in which the SiO_4^{4-} tetrahedron is the basic unit (Figure 2.24). Even though each silicon ion bonds to four oxygen ions for both states, beyond this, the structure is much more disordered and irregular for the noncrystalline structure. Whether a crystalline or amorphous solid forms depends on the ease with which a random atomic structure in the liquid can transform to an ordered state during solidification. Amorphous materials, therefore, are characterized by atomic or molecular struc-

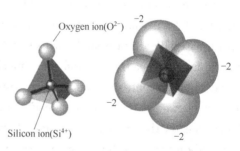

Figure 2.24 A silicon—oxygen (SiO_4^{4-}) tetrahedron

tures that are relatively complex and become ordered only with some difficulty. Furthermore, rapidly cooling through the freezing temperature favors the formation of a noncrystalline solid, since little time is allowed for the ordering process.

Metals normally form crystalline solids; but some ceramic materials are crystalline, whereas others (i.e., the silica glasses) are amorphous. Polymers may be completely noncrystalline and semicrystalline consisting of varying degrees of crystallinity. More about the structure and properties of these amorphous materials is discussed below and in subsequent chapters.

Silica Glasses

Silicon dioxide (or silica, SiO_2) in the noncrystalline state is called *fused silica*, or *vitreous silica*; again, a schematic representation of its structure is shown in Figure 2.23(b). Other oxides (e.g., B_2O_3 and GeO_2) may also form glassy structures; these materials, as well as SiO_2, are *network formers*.

The common inorganic glasses that are used for containers, windows, and so on are silica glasses to which have been added other oxides such as CaO and Na_2O. These oxides

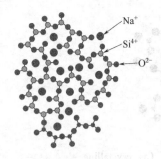

Figure 2.25 Schematic representation of ion positions in a sodium-silicate glass

do not form polyhedral networks. Rather, their cations are incorporated within and modify the SiO_4^{4-} network; for this reason, these oxide additives are termed *network modifiers*. For example, Figure 2.25 is a schematic representation of the structure of a sodium-silicate glass. Still other oxides, such as TiO_2 and Al_2O_3, while not network formers, substitute for silicon and become part of and stabilize the network; these are called *intermediates*. From a practical perspective, the addition of these modifiers and intermediates lowers the melting point and viscosity of a glass, and makes it easier to form at lower temperatures.

2.3 Imperfections in Solids

For a crystalline solid we have tacitly assumed that perfect order exists throughout the material on an atomic scale. However, such an idealized solid does not exist; all contain large numbers of various defects or imperfections. As a matter of fact, many of the properties of materials are profoundly sensitive to deviations from crystalline perfection; the influence is not always adverse, and often specific characteristics are deliberately fashioned by the introduction of controlled amounts or numbers of particular defects, as detailed in succeeding chapters. By "crystalline defect" is meant a lattice irregularity having one or more of its dimensions on the order of an atomic diameter. Classification of crystalline imperfections is frequently made according to geometry or dimensionality of the defect. Several different imperfections are discussed in this chapter, including point defects (those associated with one or two atomic positions), linear (or one-dimensional) defects, as well as interfacial defects, or boundaries, which are two-dimensional. Impurities in solids are also discussed, since impurity atoms may exist as point defects.

2.3.1 Point Defects in Metals

The simplest of the point defects is a **vacancy**, or vacant lattice site, one normally occupied from which an atom is missing (Figure 2.26). All crystalline solids contain vacancies and, in fact, it is not possible to create such a material that is free of these defects. The necessity of the existence of vacancies is explained using principles of thermodynamics; in essence, the presence of vacancies increases the entropy (i.e., the randomness) of the crystal.

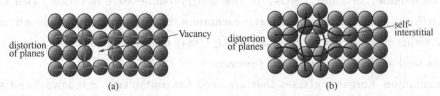

Figure 2.26 Two-dimensional representations of (a) a vacancy and (b) a self-interstitial

The equilibrium number of vacancies N_v for a given quantity of material depends on and increases with temperature according to

$$N_v = N\exp\left(-\frac{Q_v}{kT}\right) \tag{2.13}$$

In this expression, N is the total number of atomic sites, Q_v is the energy required for the formation of a vacancy, T is the absolute temperature1 in kelvins, and k is the gas or **Boltzmann's constant.** The value of k is 1.38×10^{-23} J/(atom · K), or 8.62×10^{-5} eV/(atom · K), depending on the units of Q_v. Thus, the number of vacancies increases exponentially with temperature; that is, as T in Equation 2.13 increases, so does also the expression $\exp-(Q_v/kT)$. For most metals, the fraction of vacancies N_v/N just below the melting temperature is on the order of 10^{-4}; that is, one lattice site out of 10000 will be empty. As ensuing discussions indicate, a number of other material parameters have an exponential dependence on temperature similar to that of Equation (2.13).

A **self-interstitial** is an atom from the crystal that is crowded into an interstitial site, a small void space that under ordinary circumstances is not occupied. This kind of defect is also represented in Figure 2.26. In metals, a self-interstitial introduces relatively large distortions in the surrounding lattice because the atom is substantially larger than the interstitial position in which it is situated. Consequently, the formation of this defect is not highly probable, and it exists in very small concentrations, which are significantly lower than for vacancies.

Example Problem 2.8

Calculate the equilibrium number of vacancies per cubic meter for copper at 1000℃. The energy for vacancy formation is 0.9 eV/atom; the atomic weight and density (at 1000℃) for copper are 63.5g/mol and 8.40g/cm³, respectively.

Solution

This problem may be solved by using Equation (2.13); it is first necessary, however, to determine the value of N, the number of atomic sites per cubic meter for copper, from its atomic weight A_{Cu}, its density ρ, and Avogadro's number N_A, according to

$$N = \frac{N_A \rho}{A_{Cu}}$$

$$= \frac{(6.023 \times 10^{23}\,\text{atoms/mol})(8.40\,\text{g/cm}^3)(10^6\,\text{cm}^3/\text{m}^3)}{63.5\,\text{g/mol}}$$

$$= 8.0 \times 10^{28}\,\text{atoms/m}^3$$

Thus, the number of vacancies at 1000℃ (1273 K) is equal to

$$N_v = N\exp\left(-\frac{Q_v}{kT}\right)$$

$$= (8.0 \times 10^{28}\,\text{atoms/m}^3)\exp\left[-\frac{(0.9\,\text{eV})}{(8.62 \times 10^{-5}\,\text{eV/K})(1273\text{K})}\right]$$

$$= 2.2 \times 10^{25}\,\text{vacancies/m}^3$$

Impurities in Metals

A pure metal consisting of only one type of atom just isn't possible; impurity or foreign atoms will always be present, and some will exist as crystalline point defects. In fact, even with relatively sophisticated techniques, it is difficult to refine metals to a purity in excess of 99.9999%. At this level, on the order of 10^{22} to 10^{23} impurity atoms will be present in one cubic meter of material. Most familiar metals are not highly pure; rather, they are **alloys**, in which impurity atoms have been added intentionally to impart specific characteristics to the material. Ordinarily alloying is used in metals to improve mechanical strength and corrosion resistance. For example, sterling silver is a 92.5% silver-7.5% copper alloy. In normal ambient environments, pure silver is highly corrosion resistant, but also very soft. Alloying with copper enhances the mechanical strength significantly, without depreciating the corrosion resistance appreciably.

The addition of impurity atoms to a metal will result in the formation of a **solid solution** and/or a new *second phase*, depending on the kinds of impurity, their concentrations, and the temperature of the alloy. The present discussion is concerned with the notion of a solid solution.

Several terms relating to impurities and solid solutions deserve mention. With regard to alloys, **solute** and **solvent** are terms that are commonly employed. "Solvent" represents the element or compound that is present in the greatest amount; on occasion, solvent atoms are also called *host atoms*. "Solute" is used to denote an element or compound present in a minor concentration.

Solid Solutions

A solid solution forms when, as the solute atoms are added to the host material, the crystal structure is maintained, and no new structures are formed. Perhaps it is useful to draw an analogy with a liquid solution. If two liquids, soluble in each other (such as water and alcohol) are combined, a liquid solution is produced as the molecules intermix, and its composition is homogeneous throughout. A solid solution is also compositionally homogeneous; the impurity atoms are randomly and uniformly dispersed within the solid.

Impurity point defects are found in solid solutions, of which there are two types: **substitutional** and **interstitial**. For substitutional, solute or impurity atoms replace or substitute for the host atoms (Figure 2.27). There are several features of the solute and solvent atoms that determine the degree to which the former dissolves in the latter; these are as follows:

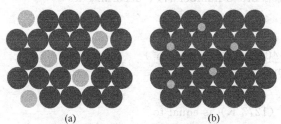

Figure 2.27 Two-dimensional schematic representations of (a) substitutional and (b) interstitial impurity atoms

1. *Atomic size factor*. Appreciable quantities of a solute may be accommodated in this type of solid solution only when the difference in atomic radii between the two atom types is less than about ±15%. Otherwise the solute atoms will create substantial lattice distortions and a new phase will form.

2. *Crystal structure*. For appreciable solid solubility the crystal structures for metals of both atom types must be the same.

3. *Electronegativity*. The more electropositive one element and the more electronegative the other, the greater is the likelihood that they will form an intermetallic compound instead of a substitutional solid solution.

4. *Valences*. Other factors being equal, a metal will have more of a tendency to dissolve another metal of higher valency than one of a lower valency.

An example of a substitutional solid solution is found for copper and nickel. These two elements are completely soluble in one another at all proportions. With regard to the aforementioned rules that govern degree of solubility, the atomic radii for copper and nickel are 0.128 and 0.125 nm, respectively, both have the FCC crystal structure, and their electronegativities are 1.9 and 1.8; finally, the most common valences are $+1$ for copper (although it sometimes can be $+2$) and $+2$ for nickel.

For interstitial solid solutions, impurity atoms fill the voids or interstices among the host atoms (see Figure 2.27). For metallic materials that have relatively high atomic packing factors, these interstitial positions are relatively small. Consequently, the atomic diameter of an interstitial impurity must be substantially smaller than that of the host atoms. Normally, the maximum allowable concentration of interstitial impurity atoms is low (less than 10%). Even very small impurity atoms are ordinarily larger than the interstitial sites, and as a consequence they introduce some lattice strains on the adjacent host atoms.

Carbon forms an interstitial solid solution when added to iron; the maximum concentration of carbon is about 2%. The atomic radius of the carbon atom is much less than that for iron: 0.071nm versus 0.124nm.

2.3.2 Dislocations—Linear Defects

A *dislocation* is a linear or one-dimensional defect around which some of the atoms are misaligned. One type of dislocation is represented in Figure 2.28: an extra portion of a plane of atoms, or half-plane, the edge of which terminates within the crystal. This is termed an **edge dislocation**; it is a linear defect that centers around the line that is defined along the end of the extra half-plane of atoms. This is sometimes termed the **dislocation line**, which, for the edge dislocation in Figure 2.28, is perpendicular to the plane of the page. Within the region around the dislocation line there is some localized lattice distortion. The atoms above the dislocation line in Figure 2.28 are squeezed together, and those below are pulled apart; this is reflected in the slight curvature for the vertical planes of atoms as they bend around this extra half-plane. The magnitude of this distortion decreases with distance away from the dislocation line; at positions far removed, the crystal lattice is virtually perfect. Sometimes the edge dislocation in Figure 2.28 is represented by the symbol-, which also indicates the position of the dislocation line. An edge dislocation may also be formed by an extra half-plane of atoms that is included in the bottom portion of

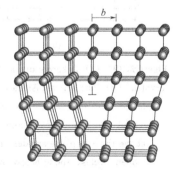

Figure 2.28 The atom positions around an edge dislocation; extra half-plane of atoms shown in perspective

the crystal; its designation is a⊥.

Another type of dislocation, called a **screw dislocation**, exists, which may be thought of as being formed by a shear stress that is applied to produce the distortion shown in Figure 2.29(a): the upper front region of the crystal is shifted one atomic distance to the right relative to the bottom portion. The atomic distortion associated with a screw dislocation is also linear and along a dislocation line, line AB in Figure 2.29(b). The screw dislocation derives its name from the spiral or helical path or ramp that is traced around the dislocation line by the atomic planes of atoms. Sometimes the symbol ⌒ is used to designate a screw dislocation.

Most dislocations found in crystalline materials are probably neither pure edge nor pure screw, but exhibit components of both types; these are termed **mixed dislocations**. The lattice distortion that is produced away from the two faces is mixed, having varying degrees of screw and edge character.

Virtually all crystalline materials contain some dislocations that were introduced during solidification, during plastic deformation, and as a consequence of thermal stresses that result from rapid cooling. Dislocations have been observed in polymeric materials; however, some controversy exists as to the nature of dislocation structures in polymers and the mechanism (s) by which polymers plastically deform.

2.3.3 Interfacial Defects

Interfacial defects are boundaries that have two dimensions and normally separate regions of the materials that have different crystal structures and/or crystallographic orientations. These imperfections include external surfaces, grain boundaries, twin boundaries, stacking faults, and phase boundaries.

2.3.3.1 External Surfaces

One of the most obvious boundaries is the external surface, along which the crystal structure terminates. Surface atoms are not bonded to the maximum number of nearest neighbors, and are therefore in a higher energy

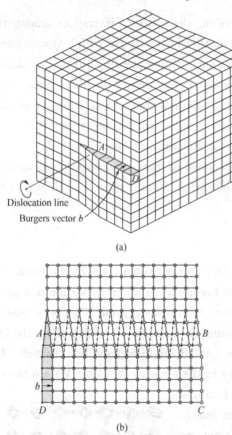

Figure 2.29 (a) A screw dislocation within a crystal. (b) The screw dislocation in (a) as viewed from above. The dislocation line extends along line AB. Atom positions above the slip plane are designated by open circles, those below by solid circles. [Figure (b) from W. T. Read, Jr., *Dislocations in Crystals*, McGraw-Hill Book Company, New York, 1953.]

state than the atoms at interior positions. The bonds of these surface atoms that are not satisfied give rise to a surface energy, expressed in units of energy per unit area (J/m^2 or erg/cm^2). To reduce this energy, materials tend to minimize, if at all possible, the total surface area. For example, liquids assume a shape having a minimum area—the droplets become spherical. Of course, this is not possible with solids, which are mechanically rigid.

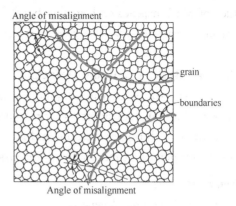

Figure 2.30 Schematic diagram showing low and high-angle grain boundaries and the adjacent atom positions

2.3.3.2 Grain Boundaries

Another interfacial defect, the grain boundary, was introduced as the boundary separating two small grains or crystals having different crystallographic orientations in polycrystalline materials. A grain boundary is represented schematically from an atomic perspective in Figure 2.30. Within the boundary region, which is probably just several atom distances wide, there is some atomic mismatch in a transition from the crystalline orientation of one grain to that of an adjacent one.

In spite of this disordered arrangement of atoms and lack of regular bonding along grain boundaries, a polycrystalline material is still very strong; cohesive forces within and across the boundary are present. Furthermore, the density of a polycrystalline specimen is virtually identical to that of a single crystal of the same material.

2.3.3.3 Twin Boundaries

A *twin boundary* is a special type of grain boundary across which there is a specific mirror lattice symmetry; that is, atoms on one side of the boundary are located in mirror image positions of the atoms on the other side (Figure 2.31). The region of material between these boundaries is appropriately termed a *twin*. Twins result from atomic displacements that are produced from applied mechanical shear forces (mechanical twins), and also during annealing heat treatments following deformation (annealing twins). Twinning occurs on a definite crystallographic plane and in a specific direction, both of which depend on the crystal structure. Annealing twins are typically found in metals that have the FCC crystal structure, while mechanical twins are observed in BCC and HCP metals.

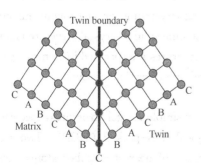

Figure 2.31 Schematic diagram showing a twin plane or boundary and the adjacent atom positions (dark circles)

2.3.3.4 Miscellaneous Interfacial Defects

Other possible interfacial defects include stacking faults, phase boundaries, and ferromagnetic domain walls. Stacking faults are found in FCC metals when there is an interruption in the ABCABCABC... stacking sequence of close-packed planes. Phase boundaries exist in multiphase materials across which there is a sudden change in physical and/or chemical characteristics.

Associated with each of the defects discussed in this section is an interfacial energy, the magnitude of which depends on boundary type, and which will vary from material to material. Normally, the interfacial energy will be greatest for external surfaces and least for domain walls.

2.3.4 Bulk or Volume Defects

Other defects exist in all solid materials that are much larger than those heretofore discussed. These include pores, cracks, foreign inclusions, and other phases. They are normally introduced during processing and fabrication steps. Some of these defects and their effects on the properties of materials are discussed in subsequent chapters.

Summary

Bonding Forces and Energies
Primary Interatomic Bonds

This chapter began with a survey of the fundamentals of atomic structure.

Atomic bonding in solids may be considered in terms of attractive and repulsive forces and energies. The three types of primary bond in solids are ionic, covalent, and metallic. For ionic bonds, electrically charged ions are formed by the transference of valence electrons from one atom type to another; forces are coulombic. There is a sharing of valence electrons between adjacent atoms when bonding is covalent. With metallic bonding, the valence electrons form a "sea of electrons" that is uniformly dispersed around the metal ion cores and acts as a form of glue for them.

Secondary Bonding or van der Waals Bonding

Both van der Waals and hydrogen bonds are termed secondary, being weak in comparison to the primary ones. They result from attractive forces between electric dipoles, of which there are two types—induced and permanent. For the hydrogen bond, highly polar molecules form when hydrogen covalently bonds to a nonmetallic element such as fluorine.

Crystal Structures
Fundamental Concepts
Unit Cells

Atoms in crystalline solids are positioned in an orderly and repeated pattern that is in contrast to the random and disordered atomic distribution found in noncrystalline or amorphous materials. Atoms may be represented as solid spheres, and, for crystalline solids, crystal structure is just the spatial arrangement of these spheres. The various crystal structures are specified in terms of parallelepiped unit cells, which are characterized by geometry and atom positions within.

Metallic Crystal Structures

Most common metals exist in at least one of three relatively simple crystal structures: face-centered cubic (FCC), body-centered cubic (BCC), and hexagonal close-packed (HCP). Two features of a crystal structure are coordination number (or number of nearest-neighbor atoms) and atomic packing factor (the fraction of solid sphere volume in the unit cell). Coordination number and atomic packing factor are the same for both FCC and HCP crystal structures, each of which may be generated by the stacking of close-packed planes of atoms.

Point Coordinates
Crystallographic Directions
Crystallographic Planes

Crystallographic points, directions, and planes are specified in terms of indexing schemes. The basis for the determination of each index is a coordinate axis system defined by the unit cell for the particular crystal structure. The location of a point within a unit cell is specified using coordinates that are fractional multiples of the cell edge lengths. Directional indices are computed in terms of the vector projection on each of the coordinate axes, whereas planar indices are determined from the reciprocals of axial intercepts.

The atomic packing (i.e., planar density) of spheres in a crystallographic plane depends on the indices of the plane as well as the crystal structure. For a given crystal structure, planes having identical atomic packing yet different Miller indices belong to the same family.

Single Crystals
Polycrystalline Materials

Single crystals are materials in which the atomic order extends uninterrupted over the entirety of the specimen; under some circumstances, they may have flat faces and regular geometric shapes. The vast majority of crystalline solids, however, are polycrystalline, being composed of many small crystals or grains having different crystallographic orientations.

Crystal Systems

Other concepts introduced in this chapter were: crystal system (a classification scheme for crystal structures on the basis of unit cell geometry); polymorphism (or allotropy) (when a specific material can have more than one crystal structure); and anisotropy (the directionality dependence of properties).

Imperfections in Solids
Vacancies and Self-Interstitials

All solid materials contain large numbers of imperfections or deviations from crystalline perfection. The several types of imperfection are categorized on the basis of their geometry and size. Point defects are those associated with one or two atomic positions; in metals these include vacancies (or vacant lattice sites), self-interstitials (host atoms that occupy interstitial sites), and impurity atoms.

Impurities in Solids

A solid solution may form when impurity atoms are added to a solid, in which case the original crystal structure is retained and no new phases are formed. For substitutional solid solutions, impurity atoms substitute for host atoms, and appreciable solubility is possible only when atomic diameters and electronegativities for both atom types are similar, when both elements have the same crystal structure, and when the impurity atoms have a valence that is the same as or less than the host material. Interstitial solid solutions form for relatively small impurity atoms that occupy interstitial sites among the host atoms.

Dislocations-Linear Defects

Dislocations are one-dimensional crystalline defects of which there are two pure types: edge and screw. An edge may be thought of in terms of the lattice distortion along the end of an extra half-plane of atoms; a screw, as a helical planar ramp. For mixed dislocations, components of both pure edge and screw are found. The magnitude and direction of lattice distortion associated with a dislocation is specified by its Burgers vector. The relative orientations of Burgers vector and dislocation line are ① perpendicular for edge, ② parallel for screw, and ③ neither perpendicular nor parallel for mixed.

Interfacial Defects

Bulk or Volume Defects

Other imperfections include interfacial defects [external surfaces, grain boundaries (both small- and high-angle), twin boundaries, etc.], volume defects (cracks, pores, etc.), and atomic vibrations. Each type of imperfection has some influence on the properties of a material.

Important Terms and Concepts

Atomic Number	Atomic Weight	Mass Unit(Amu)
Atomic Bonding	Ionic Bonding	Coulombic Force
Covalent Bonding	Metallic Bonding	Van Der Waals Bonding
Hydrogen Bonding	Dipole	Isotope
Unit Cell	Crystal System	Crystal Structure
Allotropy	Crystalline	Lattice
Amorphous	Diffraction	Lattice Parameters
Anisotropy	Face-Centered Cubic(Fcc)	Miller Indices
Atomic Packing Factor(APF)	Grain	Noncrystalline
Body-Centered Cubic(Bcc)	Grain Boundary	Polycrystalline
Bragg's Law	Hexagonal Close-Packed(Hcp)	Polymorphism
Coordination Number	Isotropic	Single Crystal
Imperfection	Point Defect	Vacancy
Boltzmann's Constant	Atom Percent	Weight Percent
Self-Interstitial	Interstitial Solid Solution	Substitutional Solid Solution
Solute	Solvent	Atomic Vibration
Dislocation Line	Screw Dislocation	Edge Dislocation
Mixed Dislocation	Burgers Vector	Grain Size

References

[1] Brady, J. E., and F. Senese, *Chemistry: Matter and Its Changes*, 4th edition, John Wiley & Sons, Inc., Hoboken, NJ, 2004.

[2] Ebbing, D. D., S. D. Gammon, and R. O. Ragsdale, *Essentials of General Chemistry*, 2nd edition, Houghton Mifflin Company, Boston, 2006.

[3] Azaroff, L. F., *Elements of X-Ray Crystallography*, McGraw-Hill, New York, 1968. Reprinted by TechBooks, Marietta, OH, 1990.

[4] Buerger, M. J., *Elementary Crystallography*, Wiley, New York, 1956.

[5] Cullity, B. D., and S. R. Stock, *Elements of X-Ray Diffraction*, 3rd edition, Prentice Hall, Upper Saddle River, NJ, 2001.

[6] *ASM Handbook*, Vol. 9, *Metallography and Microstructures*, ASM International, Materials Park, OH, 2004.

[7] Brandon, D., and W. D. Kaplan, *Microstructural Characterization of Materials*, Wiley, New York, 1999.

[8] DeHoff, R. T., and F. N. Rhines, *Quantitative Microscopy*, TechBooks, Marietta, OH, 1991.

[9] Van Bueren, H. G., *Imperfections in Crystals*, North-Holland, Amsterdam (Wiley-Interscience, New York), 1960.

[10] Vander Voort, G. F., *Metallography, Principles and Practice*, ASM International, Materials Park, OH, 1984.

[11] William D. Callister Jr., *Fundamentals of Materials Science and Engineering*, 7th Edition, John Wiley & Sons, Inc., 2007.

Chapter 3 Polymer Materials

Learning Objectives
After careful study of this chapter you should be able to do the following:
1. *Describe a typical polymer molecule in terms of its chain structure and, in addition, how the molecule may be generated from repeat units.*
2. *Calculate number—average and weight—average molecular weights, and degree of polymerization for a specified polymer.*
3. *Name and briefly describe the four general types of polymer molecular structures.*
4. *Cite the differences in behavior and molecular structure for thermoplastic and thermosetting polymers.*
5. *Briefly describe the crystalline state in polymeric materials.*
6. *Make schematic plots of the three characteristic stress—strain behaviors observed for polymeric materials.*
7. *Describe/sketch the various stages in the elastic and plastic deformations of a semicrystalline (spherulitic) polymer.*
8. *List four characteristics or structural components of a polymer that affect both its melting and glass-transition temperatures.*
9. *Cite the seven different polymer application types and, for each, note its general characteristics.*
10. *Name the five types of polymer additives and, for each, indicate how it modifies the properties.*
11. *Name and briefly describe five fabrication techniques used for plastic polymers.*

3.1 Polymer Structures

3.1.1 Introduction

Natural polymeric materials such as shellac, amber, and natural rubber have been used for centuries. A variety of other natural polymers exist, such as wood, rubber, cotton, wool, leather, and silk. Proteins, DNA, RNA and enzymes are also polymers found in plants and animals. Many of our useful plastics, rubbers, fiber and coating materials are synthetic polymers. We should care polymers since we are encircled by polymers everyday and everywhere we go. This chapter explores molecular and crystal structures of polymers.

3.1.2 Fundamental Concepts
Polymer, Mer, Monomer
　　Polymer is something made of many mers. "Mer" originates from the Greek word *mer-*

os. A mer is a basic unit that is made of carbon, hydrogen, oxygen, and/or silicon.

A monomer is a molecules that have double or triple covalent bonds that are termed unsaturated. It is possible for another atom or group of atoms to become attached to the original molecule.

3.1.3 Polymer Molecules

Most polymers are derived from unsaturated hydrocarbon termed as alkene or alkyne. The molecules in polymers are gigantic in comparison to the hydrocarbon molecules; because of their size they are often referred to as macromolecules.

A typical monomer molecule is ethylene that has a double bond between 2 carbon atoms. As the carbon atoms are unsaturated, many other ethylene molecules are possible to become attached and form a very large molecule, which is termed polyethylene (PE). The reaction from monomer ethylene to polymer polyethylene is termed polymerization.

Within each macromolecule, the mer units are bound together by covalent bonds and the backbone consisting of a string of carbon atoms is formed, represented schematically in two dimensions as follows:

$$-\overset{|}{\underset{|}{C}}-\overset{|}{\underset{|}{C}}-\overset{|}{\underset{|}{C}}-\overset{|}{\underset{|}{C}}-\overset{|}{\underset{|}{C}}-\overset{|}{\underset{|}{C}}-\overset{|}{\underset{|}{C}}-$$

Substituents are attached to each carbon atom as side groups along the backbone. Of course, a substituent can be as simple as only one hydrogen atom, or as complex as a side chain composed of many atoms.

3.1.4 Designation of Polymers

If the polymer is synthesized by only one type of monomers, the resulting polymer is called a **homopolymer**. Otherwise the polymer is called a **copolymer**.

Polymer chemists and scientists are continually searching for new materials that can be easily and economically synthesized and fabricated, with improved properties or better property combinations than are offered by the homopolymers heretofore discussed. One group of these materials are the copolymers.

Consider a copolymer that is composed of two mer units as represented by ● and in Figure 3.1. Depending on the polymerization process and the relative fractions of these mer types, different sequencing arrangements along the polymer chains are possible. For one, as depicted in Figure 3.1 (a), the two different units are randomly dispersed along the chain in what is termed a **random copolymer**. For an **alternating copolymer**, as the name suggests, the two mer units alternate chain positions, as illustrated in Figure 3.1 (b). A **block copolymer** is one in which identical mers are clustered in blocks along the chain [Figure 3.1 (c)]. And, finally, homopolymer side branches of one type may be grafted to homopolymer main chains that are composed of a different mer; such a material is termed a **graft copolymer** [Figure 3.1 (d)].

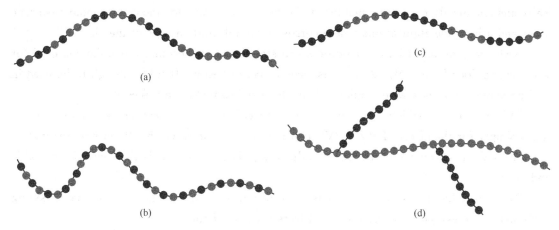

Figure 3.1 Schematic representations of (a) random, (b) alternating, (c) block, and (d) graft copolymers. The two different mer types are designated by black and colored circles. (Adapted from William D. Callister Jr., *Fundamentals of Materials Science and Engineering*, 7th Edition, John Wiley & Sons, Inc., 2007, p. 508.)

3.1.5 Commonly Used Polymers

Poly ethylene (PE)

Polyethylene is a homopolymer made from ethylene and is the most widely used plastic, with an annual production of approximately 80 million metric tons. Its primary use is within packaging (plastic bag, etc.) since it is nontoxic, smell-less and resistant to chemicals.

Polypro pylene (PP)

Polyethylene is a homopolymer made from propylene used in a wide variety of applications including packaging, textiles (e.g. ropes, thermal underwear and carpets), stationery, plastic parts and reusable containers of various types, laboratory equipment, loudspeakers, automotive components, and polymer banknotes. In 2007, the global market for polypropylene had a volume of 45.1 million tons, which led to a turnover of about 65 billion US-dollars (47.4 billion Euro).

Poly vinyl chloride (PVC)

Polyvinyl chloride is a vinyl polymer constructed of repeating vinyl groups (ethenyls) having one of their hydrogens replaced with a chloride group.

Polyvinyl chloride is the third most widely produced plastic, after polyethylene and polypropylene. PVC is widely used in construction because it is cheap, durable, and easy to assemble. PVC production is expected to exceed 40 million tons by 2016.

It can be made softer and more flexible by the addition of plasticizers, the most widely used being phthalates. In this form, it is used in clothing and upholstery, and to make flexible hoses and tubing, flooring, to roofing membranes, and electrical cable insulation. It is also commonly used in figurines and in inflatable products such as waterbeds, pool toys, and inflatable structures.

PVC's intrinsic properties make it suitable for a wide variety of applications. It is biologically and chemically resistant, making it the plastic of choice for most household sewerage pipes and other pipe applications where corrosion would limit the use of metal.

With the addition of impact modifiers and stabilizers, it becomes a popular material for window and door frames. By adding plasticizers, it can become flexible enough to be used in cabling applications as a wire insulator. It is also used to make vinyl records.

PVC has become widely used in clothing, to either create a leather-like material or at times simply for the effect of PVC. PVC clothing is common in Goth, Punk and alternative fashions. PVC is cheaper than rubber, leather, and latex and so it is more widely available and worn.

PVC fabric has a sheen to it and is waterproof. It is commonly used in coats, skiing equipment, shoes, jackets, aprons, and bags because of this.

PVC is commonly used as the insulation on electric wires; the plastic used for this purpose needs to be plasticized.

In a fire, PVC-coated wires can form HCl fumes; the chlorine serves to scavenge free radicals and is the source of the material's fire retardance. While HCl fumes can also pose a health hazard in their own right, HCl dissolves in moisture and breaks down onto surfaces, particularly in areas where the air is cool enough to breathe, and is not available for inhalation. Frequently in applications where smoke is a major hazard (notably in tunnels and communal areas) PVC-free cable insulation is preferred, such as low smoke zero halogen (LSZH) insulation.

Roughly half of the world's PVC resin manufactured annually is used for producing pipes for various municipal and industrial applications. In the water distribution market it accounts for 66% of the market in the US, and in sanitary sewer pipe applications, it accounts for 75%. Its light weight, high strength, and low reactivity make it particularly well-suited to this purpose. In addition, PVC pipes can be fused together using various solvent cements, or heat-fused (butt-fusion process, similar to joining HDPE pipe), creating permanent joints that are virtually impervious to leakage.

Many vinyl products contain additional chemicals to change the chemical consistency of the product. Some of these additional chemicals called additives which may do harm to health can leach out of vinyl products.

Poly tetra fluoro ethylene (PTFE)

Polytetrafluoroethylene is a homopolymer made from tetrafluoroethylene and is most well known by the DuPont brand name Teflon. PTFE is slightly different from PE in that all the hydrogen atoms in PE are replace by fluorine atoms. Many excellent properties make PTFE find numerous applications. PTFE has one of the lowest coefficients of friction against any solid and can be used as a lubricant. It is very non-reactive, partly because of the strength of carbon-fluorine bonds, and so it is often used in containers and pipework for reactive and corrosive chemicals. PTFE is also used as a non-stick coating for pans and other cookware due to its lowest surface energy density among all solid matters.

Poly styrene (PS)

Styrene is the ethylene in which one H atom is replaced/substituted by a benzene ring. The corresponding polymer is PS which is one of the most widely used kinds of plastic. Pure solid polystyrene is a colorless, hard plastic with limited flexibility. It can be cast into molds with fine detail. Polystyrene can be transparent or can be made to take on various colors.

Solid polystyrene is used, for example, in disposable cutlery, plastic models, CD and DVD cases, and smoke detector housings. Products made from foamed polystyrene are nearly ubiquitous, for example packing materials, insulation, and foam drink cups.

Poly amide (PA)

A polyamide is a polymer containing monomers of amides joined by peptide bonds. The amide link is produced from the condensation reaction of an amino group and a carboxylic acid or acid chloride group. A small molecule, usually water, or hydrogen chloride, is eliminated.

PA can occur both naturally and artificially, examples being proteins, such as wool and silk, and can be made artificially through step-growth polymerization or solid-phase synthesis, examples being nylons, aramids, and sodium poly (aspartate). Polyamides are commonly used in textiles, automotives, carpet and sportswear due to their extreme durability and strength.

Nylon is a generic designation for a family of synthetic polymers known generically as polyamides, first produced on February 28, 1935, by Wallace Carothers at DuPont's research facility at the DuPont Experimental Station. Nylon is one of the most commonly used polymers. Nylon was intended to be a synthetic replacement for silk and substituted for it in many different products after silk became scarce during World War II. It replaced silk in military applications such as parachutes and flak vests, and was used in many types of vehicle tires.

Nylon fibres are used in many applications, including fabrics, bridal veils, carpets, musical strings, and rope.

Solid nylon is used for mechanical parts such as machine screws, gears and other low- to medium-stress components previously cast in metal. Engineering-grade nylon is processed by extrusion, casting, and injection molding. Solid nylon is used in hair combs. Type 6, 6 Nylon 101 is the most common commercial grade of nylon, and Nylon 6 is the most common commercial grade of molded nylon. Nylon is available in glass-filled variants which increase structural and impact strength and rigidity, and molybdenum sulfide-filled variants which increase lubricity.

Aramid (= aromatic polyamides) is made from two different monomers which continuously alternate to form the polymer and is an aromatic polyamide. Aramid fibers are a class of heat-resistant and strong synthetic fibers. They are used in aerospace and military applications, for ballistic rated body armor fabric and ballistic composites, in bicycle tires, and as an asbestos substitute. The name is a portmanteau of "aromatic polyamide". They are fibers in which the chain molecules are highly oriented along the fiber axis, so the strength of the chemical bond can be exploited.

Polyester

Polyester is a category of polymers which contain the ester functional group [—(C=

O)—O—] in their main chain. Although there are many polyesters, the term "polyester" as a specific material most commonly refers to polyethylene terephthalate (PET). Polyesters include naturally-occurring chemicals, such as in the cutin of plant cuticles, as well as synthetics through step-growth polymerization such as Polyethylene terephthalate, polycarbonate and polybutyrate. Natural polyesters and a few synthetic ones are biodegradable, but most synthetic polyesters are not.

1) Poly ethylene terephthalate (PET)

Polyethylene terephthalate, commonly abbreviated **PET**, **PETE**, or the obsolete PETP or PET-P, is a thermoplastic polymer resin of the polyester family and consists of polymerized units of the monomer ethylene terephthalate, with repeating $C_{10}H_8O_4$ units. PET is commonly recycled, and has the number "1" as its recycling symbol.

Depending on its processing and thermal history, polyethylene terephthalate may exist both as an amorphous (transparent) and as a semi-crystalline polymer. The semicrystalline material might appear transparent (particle size $<$ 500 m) or opaque and white (particle size up to a few microns) depending on its crystal structure and particle size.

The majority of the world's PET production is for synthetic fibers (in excess of 60%) with bottle production accounting for around 30% of global demand. PET fibers have a trademark of "Dacron". Its properties include high tensile strength, high resistance to stretching.

While synthetic clothing in general is perceived by some as having a less-natural feel compared to fabrics woven from natural fibres (such as cotton and wool), polyester fabrics can provide specific advantages over natural fabrics, such as improved wrinkle resistance. As a result, polyester fibres are sometimes spun together with natural fibres to produce a cloth with blended properties. Synthetic fibres also can create materials with superior water, wind and environmental resistance compared to plant-derived fibres.

Plastic bottles made from PET are excellent barrier materials and are widely used for soft drinks. For certain specialty bottles, PET sandwiches an additional polyvinyl alcohol to further reduce its oxygen permeability.

2) Poly carbonates (PC)

Polycarbonates received their name because they are polymers containing carbonate groups $[-O-(C=O)-O-]$. Most polycarbonates of commercial interest are derived from rigid monomers. A balance of useful features including temperature resistance, impact resistance and optical properties position polycarbonates between commodity plastics and engineering plastics.

Polycarbonate is a very durable material. Although it has high impact-resistance, it has low scratch-resistance and so a hard coating is applied to polycarbonate eyewear lenses and polycarbonate exterior automotive components. The characteristics of polycarbonate are quite like those of polymethyl methacrylate (PMMA, acrylic), but polycarbonate is stronger, usable in a wider temperature range but more expensive. This polymer is highly transparent to visible light and has better light transmission characteristics than many kinds of glass.

Polycarbonate can undergo large plastic deformations without cracking or breaking. As a result, it can be processed and formed at room temperature using sheet metal techniques, such as forming bends on a brake. Even for sharp angle bends with a tight radius, no heating is generally necessary. This makes it valuable in prototyping applications where transparent or electrically non-conductive parts are needed, which cannot be made from sheet metal.

Polycarbonate is mainly used for electronic applications that capitalize on its collective safety features. Being a good electrical insulator and having heat resistant and flame retardant properties, it is used in various products associated with electrical and telecommunications hardware. It also serves as dielectric in high stability capacitors.

The second largest consumer of polycarbonates is the construction industry, e. g. for domelights, flat or curved glazing, and sound walls.

A major application of polycarbonate is the production of compact discs, DVDs, and Blu-ray Discs. The blanks are produced by injection molding. Typical products of sheet/film production include applications in advertisement (signs, displays, poster protection).

Many kinds of lenses are manufactured from polycarbonate, including automotive headlamp lenses, lighting lenses, sunglass/eyeglass, lenses, and safety glasses. Other miscellaneous items: MP3/Digital audio player cases, Ocarinas, computer cases, riot shields, visors, instrument panels. Many toys and hobby items are made from polycarbonate parts, e. g. fins, gyro mounts, and flybar locks for use with radio-controlled helicopters.

In the automotive industry, injection moulded polycarbonate can produce very smooth surfaces that make it well suited for direct (without the need for a basecoat) metalised parts such as decorative bezels and optical reflectors. Its uniform mould shrinkage results in parts with greater accuracy than those made of polypropylene. However, due to its susceptibility to environmental stress cracking, its use is limited to low stress applications. It can be laminated to make bullet-proof "glass", although "bullet-resistant" is more accurate for the thinner windows, such as are used in bullet-resistant windows in automobiles. The thicker barriers of transparent plastic used in teller's windows and barriers in banks, are also polycarbonate.

3) Poly urethane (PU)

A polyurethane (IUPAC abbreviation PUR, but commonly abbreviated PU) is any polymer consisting of a chain of organic units joined by urethane (carbamate) links produced by reacting an isocyanate group, —N=C=O with a hydroxyl (alcohol) group, —OH.

Polyurethane products have many uses. Over three quarters of the global consumption of polyurethane products is in the form of foams, with flexible and rigid types being roughly equal in market size. In both cases, the foam is usually behind other materials: flexible foams are behind upholstery fabrics in commercial and domestic furniture; rigid foams are inside the metal and plastic walls of most refrigerators and freezers, or behind paper, metals and other surface materials in the case of thermal insulation panels in the construction sector. Its use in garments is growing: for example, in lining the cups of brassieres. Polyurethane is also used for moldings which include door frames, columns, balusters, window headers, pediments, medallions and rosettes. The new iPad Smart Cover is made of polyurethane.

Polyurethane formulations cover an extremely wide range of stiffness, hardness, and densities. These materials include:

- Low-density flexible foam used in upholstery, bedding, and automotive and

truck seating
- Low-density rigid foam used for thermal insulation and RTM cores
- Soft solid elastomers used for gel pads and print rollers
- Low density elastomers used in footwear
- Hard solid plastics used as electronic instrument bezels and structural parts
- Flexible plastics used as straps and bands

Polyurethane foam is widely used in high resiliency flexible foam seating, rigid foam insulation panels, microcellular foam seals and gaskets, durable elastomeric wheels and tires, automotive suspension bushings, electrical potting compounds, seals, gaskets, carpet underlay, and hard plastic parts (such as for electronic instruments).

Poly (methyl meth acrylate) (PMMA)

PMMA is a transparent plastic, often used as a light or shatter-resistant alternative to glass. It is sometimes called acrylic glass. Chemically, it is the synthetic polymer of methyl methacrylate.

PMMA is an economical alternative to polycarbonate (PC) when extreme strength is not necessary. Additionally, PMMA does not contain the potentially harmful bisphenol-A subunits found in polycarbonate. It is often preferred because of its moderate properties, easy handling and processing, and low cost, but behaves in a brittle manner when loaded, especially under an impact force, and is more prone to scratching compared to glass.

PMMA acrylic glass is commonly used for constructing residential and commercial aquariums. Designers started building big aquariums when PMMA could be used. The spectacular size of both flat panels and tunnels in aquariums such as Monterey Bay, Tokyo Sea Life Park, Osaka, Nagoya, Georgia and Dubai Aquariums were made possible with the introduction of acrylic. PMMA is used in the lenses of exterior lights of automobiles.

The spectator protection in ice hockey rinks is made from PMMA. It is also used in motorcycle helmet visors. Polycast acrylic sheet is the most widely used material in aircraft transparencies (windows). In applications where the aircraft is pressurized, stretched acrylic is used. Only in the most advanced modern fighter jets, such as the F-22 Raptor, has traditional acrylic been replaced by polycarbonate. In some Motor racing championships the glass windows in the cars are replaced with acrylic to prevent glass shattering on the driver and track during a crash. They also help to save some weight making the car lighter and faster.

PMMA has a good degree of compatibility with human tissue, and can be used for replacement intraocular lenses in the eye when the original lens has been removed in the treatment of cataracts. This compatibility was discovered in WW Ⅱ RAF pilots, whose eyes had been riddled with PMMA splinters coming from the side windows of their Supermarine Spitfire fighters - the plastic scarcely caused any rejection, compared to glass splinters coming from aircraft such as the Hawker Hurricane. Historically, hard contact lenses were frequently made of this material. Soft contact lenses are often made of a related polymer, where acrylate monomers containing one or more hydroxyl groups make them hydrophilic.

In orthopedic surgery, PMMA bone cement is used to affix implants and to remodel lost bone. It is supplied as a powder with liquid methyl methacrylate (MMA). When mixed these yield a dough-like cement that gradually hardens. Surgeons can judge the curing of the PMMA bone cement by pressing their thumb on it. Although PMMA is biologically compatible, MMA is considered to be an irritant and a possible carcinogen. PMMA has also been linked to cardiopulmonary events in the operating room due to hypotension. Bone cement acts like a grout and not so much like a glue in arthroplasty. Although sticky, it does not bond to either the bone or the implant, it primarily fills the spaces between the prosthesis and the bone preventing motion. A big disadvantage to this bone cement is that it heats to quite a high temperature while setting, potentially 82.5℃ and because of this, thermal necrosis of neighboring tissue can potentially result. A careful balance of initiators and monomers is needed to reduce the rate of polymerization, and thus the heat generated. A major consideration when using PMMA cement is the effect of stress shielding. Since PMMA has a Young's modulus greater than that of natural bone, the stresses are loaded into the cement and so the bone no longer receives the mechanical signals to continue bone remodeling and so resorption will occur.

Dentures are often made of PMMA, and can be color-matched to the patient's teeth & gum tissue. PMMA is also used in the production of ocular prostheses.

In cosmetic surgery, tiny PMMA microspheres suspended in some biological fluid are injected under the skin to reduce wrinkles or scars permanently.

A large majority of white Dental filling materials (i.e. composites) have PMMA as their main organic component.

Acrylic paint essentially consists of PMMA suspended in water; however since PMMA is hydrophobic, a substance with both hydrophobic and hydrophilic groups needs to be added to facilitate the suspension.

Modern furniture makers, especially in the 1960s and 1970s, seeking to give their products a space age aesthetic, incorporated Lucite and other PMMA products into their designs, especially office chairs. Many other products (for example, guitars) are sometimes made with acrylic glass to make the commonly opaque objects translucent.

In the 1950s and 1960s, Lucite was an extremely popular material for jewelry, with several companies specialized in creating high-quality pieces from this material. Lucite beads and ornaments are still sold by jewelry suppliers.

Polysiloxane (Silicones)

Polysiloxane or silicones are mixed inorganic-organic polymers that include silicon together with carbon, hydrogen, oxygen, and sometimes other chemical elements with the chemical formula $[R_2SiO]_n$, where R is an organic group such as methyl, ethyl, or phenyl. These materials consist of an inorganic silicon-oxygen backbone (····—Si—O—Si—O—Si—O—····) with organic side groups attached to the silicon atoms, which are four-coordinate. Some common forms include silicone oil, silicone grease, silicone rubber, and silicone resin.

In some cases organic side groups can be used to link two or more of these —Si—O— backbones together. By varying the —Si—O— chain lengths, side groups, and crosslinking, silicones can be synthesized with a wide variety of properties and compositions. They can vary in consistency from liquid to gel to rubber to hard plastic. The most common siloxane is linear polydimethylsiloxane (PDMS), a silicone oil. The second largest group of silicone materials is based on silicone resins, which are formed by branched and cage-like oligosiloxanes.

Some of the most useful properties of silicones can include:

- Good electrical insulation. Because silicone can be formulated to be electrically insulative or conductive, it is suitable for a wide range of electrical applications.
- Thermal stability (constancy of properties over a wide operating range of −100 to 250℃).
- Though not a hydrophobe, the ability to repel water and form watertight seals.
- Excellent resistance to oxygen, ozone and UV light (sunlight). This has led to widespread use in the construction industry (e.g. coatings, fire protection, glazing seals), and automotive industry (external gaskets, external trim).
- Does not stick.
- Low chemical reactivity.
- Low toxicity, but does not support microbiological growth.
- High gas permeability: at room temperature (25℃) the permeability of silicone rubber for gases like oxygen is approximately 400 times that of butyl rubber, making silicone useful for medical applications that could benefit from increased aeration. Silicone rubbers cannot be used where gas-tight seals are necessary.

3.1.6 The Chemistry of Polymer Molecules

Why the polymers are so different? The answer is: the properties of polymers are influenced by the following factors: molecular weight, molecular shape, molecular structure and polymer crystallinity.

3.1.6.1 Molecular weight

Extremely large molecular weights1 are to be found in polymers with very long chains. During the polymerization process in which these large macromolecules are synthesized from smaller molecules, not all polymer chains will grow to the same length; this results in a distribution of chain lengths or molecular weights. Ordinarily, an average molecular weight is specified, which may be determined by the measurement of various physical properties such as viscosity and osmotic pressure.

There are several ways of defining average molecular weight. The number average molecular weight $\overline{M}_n$ is obtained by dividing the chains into a series of size ranges and then determining the number fraction of chains within each size range [Figure 3.2 (a)]. This number-average molecular weight is expressed as

$$\overline{M}_n = \sum x_i M_i \tag{3.1a}$$

where M_i represents the mean (middle) molecular weight of size range i, and x_i is the fraction of the total number of chains within the corresponding size range.

A weight-average molecular weight $\overline{M}_w$ is based on the weight fraction of molecules

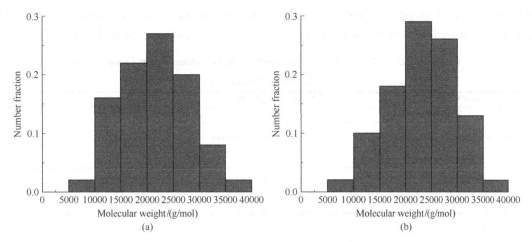

Figure 3.2 Hypothetical polymer molecule size distributions on the basis of (a) number and (b) weight fractions of molecules

within the various size ranges [Figure 3.2 (b)]. It is calculated according to

$$\overline{M}_w = \sum w_i M_i \tag{3.1b}$$

where, again, M_i is the mean molecular weight within a size range, whereas w_i enotes the weight fraction of molecules within the same size interval. Computations for both number-average and weight-average molecular weights are carried out in Example Problem 3.1. A typical molecular weight distribution along with these molecular weight averages are shown in Figure 3.3.

An alternate way of expressing average chain size of a polymer is as the **degree of polymerization** n, which represents the average number of mer units in a chain. Both number-average (n_n) and weight-average (n_w) degrees of polymerization are possible, as follows:

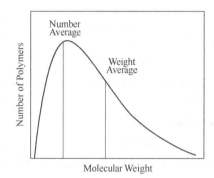

Figure 3.3 Distribution of molecular weights for a typical polymer

$$n_n = \frac{\overline{M}_n}{\overline{m}} \tag{3.2a}$$

$$n_w = \frac{\overline{M}_w}{\overline{m}} \tag{3.2b}$$

Where $\overline{M}_n$ and $\overline{M}_w$ are the number-average and weight-average molecular weights as defined above, while m is the mer molecular weight. For a copolymer (having two or more different mer units), m is determined from

$$\overline{m} = \sum f_j m_j \tag{3.3}$$

In this expression, f_j and m_j are, respectively, the chain fraction and molecular weight of mer j.

Example Problem **3.1**

Assume that the molecular weight distributions shown in Figure 3.2 are for polyvinyl chloride. For this material, compute (a) the number-average molecular weight; (b) the number-average degree of polymerization; and (c) the weight average molecular weight.

Table 3.1 (a)　　Data Used for Number-Average Molecular Weight Computations in Example Problem 3.1

Molecular Weight Range/(g/mol)	Mean M_i/(g/mol)	x_i	$x_i M_i$
5000~10000	7500	0.05	375
10000~15000	12500	0.16	2000
15000~20000	17500	0.22	3850
20000~25000	22500	0.27	6075
25000~30000	27500	0.20	5500
30000~35000	32500	0.08	2600
35000~40000	37500	0.02	750

$\overline{M}_n = 21150$.

Table 3.1 (b)　　Data Used for Weight-Average Molecular Weight Computations in Example Problem 3.1

Molecular Weight Range/(g/mol)	Mean M_i/(g/mol)	w_i	$w_i M_i$
5000-10000	7500	0.02	150
10000-15000	12500	0.10	1250
15000-20000	17500	0.18	3150
20000-25000	22500	0.29	6525
25000-30000	27500	0.26	7150
30000-35000	32500	0.13	4225
35000-40000	37500	0.02	750

$\overline{M}_w = 23200$.

Solution

(a) The data necessary for this computation, as taken from Figure 3.2, are presented in Table 3.1 (a). According to Equation (3.2), summation of all the $x_i M_i$ products (from the right-hand column) yields the number-average molecular weight, which in this case is 21150 g/mol.

(b) To determine the number-average degree of polymerization, it first becomes necessary to compute the mer molecular weight. For PVC, each mer consists of two carbon atoms, three hydrogen atoms, and a single chlorine atom (Table 3.1). Furthermore, the atomic weights of C, H, and Cl are, respectively, 12.01, 1.01, and 35.45 g/mol. Thus, for PVC

$$\overline{m} = 2(12.01 \text{ g/mol}) + 3(1.01 \text{ g/mol}) + 35.45 \text{g/mol}$$
$$= 62.50 \text{ g/mol}$$

and

$$n_n = \frac{\overline{M}_n}{\overline{m}} = \frac{21150 \text{ g/mol}}{62.50 \text{ g/mol}} = 338$$

(c) Table 3.1 (b) shows the data for the weight-average molecular weight, as taken from Figure 3.1 (b). The $w_i M_i$ products for the several size intervals are tabulated in the right-hand column. The sum of these products [Equation (3.2b)] yields a value of 23200 g/mol for $\overline{M}_w$.

Various polymer characteristics are affected by the magnitude of the molecular weight. One of these is the melting or softening temperature; melting temperature is raised with increasing molecular weight (for $\overline{M}$ up to about 100000 g/mol). At room temperature, polymers with very short chains (having molecular weights on the order of 100 g/mol) exist as liquids or gases. Those with molecular weights of approximately 1000 g/mol are waxy solids (such as paraffin wax) and soft resins. Solid polymers (sometimes termed *high polymers*), which are of prime interest here, commonly have molecular weights ranging between 10000 and several million g/mol (Table 3.2).

Table 3.2 A Listing of Mer Structures for 10 of the More Common Polymeric Materials

Polymer	Repeating(Mer)structure
Polyethylene(PE)	—CH$_2$—CH$_2$—
Polyvinyl chloride(PVC)	—CHCl—CH$_2$—
Polytetrafluoroethylene(PTFE)	—CF$_2$—CF$_2$—
Polypropylene(PP)	—CH(CH$_3$)—CH$_2$—
Polystyrene(PS)	—CH(C$_6$H$_5$)—CH$_2$—
Poly(methyl methacrylate)(PMMA)	—CH$_2$—C(CH$_3$)(COOCH$_3$)—
Polyethylene terephthalate(PET, a polyester)	—OC—C$_6$H$_4$—CO—O—CH$_2$—CH$_2$—O—
Poly(hexamethylene adipamide)(nylon 6,6)	—NH—(CH$_2$)$_6$—NH—CO—(CH$_2$)$_4$—CO—
Polycarbonate	—O—C$_6$H$_4$—C(CH$_3$)$_2$—C$_6$H$_4$—O—CO—

3.1.6.2 Molecular Shape

Consider again the polyethylene composed of many ethylene monomer units. This representation is not strictly correct in that the angle between the singly bonded carbon atoms is not 180° as shown, but rather close to 109°. A more accurate three-dimensional model is one in which the carbon atoms form a zigzag pattern (Figure 3.4), the C-C bond length being 0.154 nm. In this discussion, depiction of polymer molecules is frequently simplified using the linear chain model.

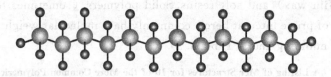

Figure 3.4 A perspective of the polyethylene molecule, indicating the zigzag backbone structure

There is no reason to suppose that polymer chain molecules are strictly straight, in the sense that the zigzag arrangement of the backbone atoms (Figure 3.4) is disregarded. Single chain bonds are capable of rotation and bending in three dimensions. Consider the chain atoms in Figure 3.5 (a); a third carbon atom may lie at any point on the cone of revolution and still subtend about a 109° angle with the bond between the other two atoms. A straight chain segment results when successive chain atoms are positioned as in Figure 3.5 (b). On the other hand, chain bending and twisting are possible when there is a rotation of the chain atoms into other positions, as illustrated in Figure 3.5 (c). Thus, a single chain molecule composed of many chain atoms might assume a shape similar to that represented schematically in Figure 3.6, having a multitude of bends, twists, and kinks. Also indicated in this figure is the end-to-end distance of the polymer chain r; this distance is much smaller than the total chain length.

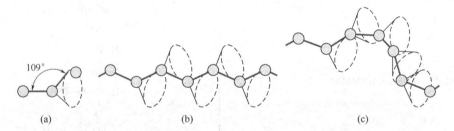

Figure 3.5 Schematic representations of how polymer chain shape is influenced by the positioning of backbone carbon atoms (solid circles). For (a), the rightmost atom may lie anywhere on the dashed circle and still subtend a 109° angle with the bond between the other two atoms. Straight and twisted chain segments are generated when the backbone atoms are situated as in (b) and (c), respectively.
(Adapted from William D. Callister Jr., *Fundamentals of Materials Science and Engineering*, 7th Edition, John Wiley & Sons, Inc., 2007, p. 501.)

Polymers consist of large numbers of molecular chains, each of which may bend, coil, and kink in the manner of Figure 3.6. This leads to extensive intertwining and entanglement of neighboring chain molecules, a situation similar to that of a fishing line that has experienced backlash from a fishing reel. These random coils and molecular entanglements are responsible for a number of important characteristics of polymers, to include the large elastic extensions displayed by the rubber materials.

Some of the mechanical and thermal characteristics of polymers are a function of the ability of chain segments to experience rotation in response to applied stresses or thermal vibrations. Rotational flexibility is dependent on mer structure and chemistry. For example, the region of a chain segment that has a double bond (C=C) is rotationally rigid. Also, introduction of a bulky or large side group of atoms restricts rotational movement. For example, polystyrene molecules, which have a phenyl side group, are more resistant to rotational motion than are polyethylene chains.

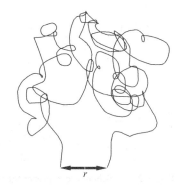

Figure 3.6 Schematic representation of a single polymer chain molecule that has numerous random kinks and coils produced by chain bond rotations

3.1.6.3 Molecular Structure

The physical characteristics of a polymer depend not only on its molecular weight and shape, but also on differences in the structure of the molecular chains. Modern polymer synthesis techniques permit considerable control over various structural possibilities. This section discusses several molecular structures including linear, branched, crosslinked, and network, in addition to various isomeric configurations.

Linear Polymers

Linear polymers are those in which the mer units are joined together end to end in single chains. These long chains are flexible and may be thought of as a mass of spaghetti, as represented schematically in Figure 3.7 (a), where each circle represents a mer unit. For linear polymers, there may be extensive van der Waals and hydrogen bonding between the chains. Some of the common polymers that form with linear structures are polyethylene, polyvinyl chloride, polystyrene, polymethyl methacrylate, nylon, and the fluorocarbons.

Branched Polymers

Polymers may be synthesized in which side-branch chains are connected to the main ones, as indicated schematically in Figure 3.7 (b); these are fittingly called **branched polymers**. The branches, considered to be part of the main-chain molecule, result from side reactions that occur during the synthesis of the polymer. The chain packing efficiency is reduced with the formation of side branches, which results in a lowering of the polymer density. Those polymers that form linear structures may also be branched.

Thermoplastic and Thermosetting Polymers

A thermoplastic, also known as a thermosoftening plastic or thermoplasts, is a polymer soften when heated (and eventually liquefy) and harden to a very glassy when cooled—processes that are totally reversible and may be repeated. A thermosetting plastic, also known as a thermoset, is polymer becomes permanently hard when crosslinking is applied and do not soften upon subsequent heating.

On a molecular level, thermoplastic polymers differ from thermosetting polymers in that, for thermoplasts only weak secondary bonding forces exist between chains whereas for

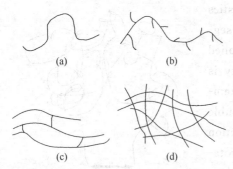

Figure 3.7 Schematic representations of (a) linear, (b) branched, (c) crosslinked, and (d) network (three-dimensional) molecular structures

thermosets chains are crosslinked by strong primary bonding forces (i.e. covalent bonds) together with secondary bonding forces between molecules. As the temperature is raised, secondary bonding forces are diminished (by increased molecular motion) so that the relative movement of adjacent chains within a thermoplast is facilitated when a stress is applied. Irreversible degradation results when the temperature of a molten thermoplastic polymer is raised to the point at which molecular vibrations become violent enough to break the primary covalent bonds. For a thermoset, crosslinking formed by strong primary bonding forces between chains cannot be destroyed unless heating to excessive temperatures, which will cause severance of these crosslink bonds as well as polymer degradation. Consequently thermoplasts are recyclable but thermosets are irrecyclable.

In addition, thermoplasts are relatively soft and soluble in proper solvents. Most linear polymers and those having some branched structures with flexible chains are thermoplastic. These materials are normally fabricated by the simultaneous application of heat and pressure. Many thermoplastic materials are addition polymers; e.g., vinyl chain-growth polymers such as polyethylene and polypropylene. Thermoset polymers are generally harder and stronger than thermoplastics, and have better dimensional stability. Most of the crosslinked and network polymers are not soluble, which include vulcanized rubbers, epoxies, and phenolic and some polyester resins, are thermosetting.

3.1.6.4 Polymer Crystallinity

Polymer crystallinity defines the regular packing of molecular chains that produces an ordered atomic array. The atomic arrangements will be more complex for polymers, since it involves molecules instead of just atoms or ions, as with metals and ceramics. Twisting, kinking and coiling of the chains also prevent the chains from packing orderly. As a consequence, polymer molecules are often only partially crystalline (or semicrystalline), having crystalline regions dispersed within the remaining amorphous material. Within the ordered regions, the polymer chains are both aligned and folded and form what called lamellae, which compose larger spheroidal structures named spherulites (Figure 3.8).

Whether or not polymers can crystallize depends on their molecular structure — simple mer units, regularly spaced smaller side groups and less cross-

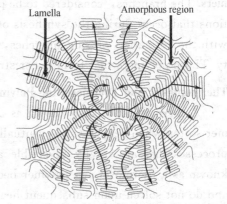

Figure 3.8 Schematic representation of a polymer spherulite

linked mer units facilitates crystallization. For example, crystallization occurs very much easier in PE and PTFE that having simplest mer units and smallest side groups than in PC that have complex mer units, or than rubber or silicones that are highly crosslinked.

For copolymers, as a general rule, those with regular arrangement of mer types have the greater tendency to be crystallized. In fact, alternating and block copolymers may be partially crystalline; random and branched copolymers are normally amorphous.

The degree of crystallinity of a polymer also depends on the rate of cooling during solidification as well as on the chain configuration. During crystallization upon cooling through the melting temperature, the chains, which are highly random and entangled in the viscous liquid, must assume an ordered configuration. For this to occur, sufficient time must be allowed for the chains to move and align themselves.

The degree of crystallinity of a polymer, defined as the fraction of the ordered molecules in polymer, may range from completely amorphous to almost entirely (up to about 95% for PE and PTFE) crystalline; by way of contrast, metal specimens are almost always entirely crystalline, whereas many ceramics are either totally crystalline or totally noncrystalline. Semicrystalline polymers are, in a sense, analogous to two phase metal alloys, discussed in subsequent chapters.

The density of a crystalline polymer will be greater than an amorphous one of the same material and molecular weight, since the chains are more closely packed together for the crystalline structure. The degree of crystallinity by weight may be determined from accurate density measurements, according to

$$\% \text{ crystallinity} = \frac{\rho_c(\rho_s-\rho_a)}{\rho_s(\rho_c-\rho_a)} \times 100\% \tag{3.4}$$

where ρ_s is the density of a specimen for which the percent crystallinity is to be determined, ρ_a is the density of the totally amorphous polymer, and ρ_c is the density of the perfectly crystalline polymer. The values of ρ_a and ρ_c must be measured by other experimental means.

To some extent, the physical properties of polymeric materials are influenced by the degree of crystallinity. Crystalline polymers are usually stronger and more resistant to dissolution and softening by heat.

3.2 Crystallization, Melting and Glass Transition Phenomena in Polymers

Three phenomena that are important with respect to the design and processing of polymeric materials are crystallization, melting, and the glass transition. Crystallization is the process by which, upon cooling, an ordered (i.e., crystalline) solid phase is produced from a liquid melt having a highly random molecular structure. The melting transformation is the reverse process that occurs when a polymer is heated.

Crystallization and melting are only possible for crystalline polymers. Below melting temperature (T_m), the crystalline polymer is solid whereas liquid above T_m. In amorphous polymers, things are not so straightforward. An amorphous polymer does not have a phase transition temperature (such as T_m) but have a glass transition temperature (T_g). The word "glass" indicates that upon heating an amorphous polymer is just like a glass: below T_g, an amorphous polymer is solid, yet has no long-range molecular order and so is noncrystalline; the material is viscous liquid when it is held at temperatures above T_g. Upon cooling, the glass transition corresponds to the gradual transformation from a liquid to a rubbery material, and finally, to a rigid solid.

Many polymers are semicrystalline having crystalline regions dispersed within the remaining amorphous material. Consequently, they have both T_m and T_g, i.e., crystalline regions will experience melting (and crystallization), while noncrystalline areas pass through the glass transition.

Melting and glass transition temperatures are important parameters relative to inservice applications of polymers. They define, respectively, the upper and lower temperature limits for numerous applications, especially for semicrystalline polymers. The glass transition temperature may also define the upper use temperature for glassy amorphous materials. Furthermore, and also influence the fabrication and processing procedures for polymers and polymer-matrix composites. These issues are discussed in succeeding sections of this chapter. Representative melting and glass transition temperatures of a number of polymers are contained in Table 3.3.

Table 3.3 Melting and Glass Transition Temperatures for Some of the More Common Polymeric Materials

Polymeric Materials	Glass Transition Temperature/[℃(℉)]	Melting Temperature/(℃/℉)
PE(low density)	−110(−165)	115(240)
PTFE	−97(−140)	327(620)
PE(high density)	−90(−130)	137(279)
PP	−18(0)	175(347)
Nylon 6,6	57(135)	265(510)
Poly	69(155)	265(510)
PVC	87(190)	212(415)
PS	100(212)	240(465)
PC	150(300)	265(510)

3.3 Mechanical Properties of Polymers

Polymers are used in a wide variety of applications from construction materials to microelectronics processing. Thus, most engineers will be required to work with polymers at some point in their careers. Understanding the mechanisms by which polymers elastically and plastically deform allows one to alter and control their moduli of elasticity and strengths. Also, additives may be incorporated into polymeric materials to modify a host of properties, including strength, abrasion resistance, toughness, thermal stability, stiffness, deteriorability, color, and flammability resistance.

3.3.1 Stress-Strain Behavior

Stress-strain curves show the response of a material to an applied (usually tensile) stress. They allow important information such as a material's modulus of elasticity and yield and tensile strengths to be determined. Accurate knowledge of these parameters is paramount in engineering design. Polymer stress-strain curves are produced by stretching a sample at a constant rate through the application of a tensile force.

The mechanical characteristics of polymers, for the most part, are highly sensitive to the rate of deformation (strain rate), the temperature, and the chemical nature of the environment (the presence of water, oxygen, organic solvents, etc.). Some modifications of the testing techniques and specimen configurations used for metals are necessary with polymers, especially for the highly elastic materials, such as rubbers.

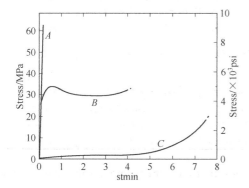

Figure 3.9 The stress—strain behavior for brittle (curve A), plastic (curve B), and highly elastic (elastomeric) (curve C) polymers

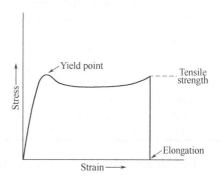

Figure 3.10 Schematic stress-strain curve for a plastic polymer showing how yield and tensile strengths are determined

Three typically different types of stress—strain behavior are found for polymeric materials, as represented in Figure 3.9. Curve A illustrates the stress—strain character for a brittle polymer, inasmuch as it fractures while deforming elastically. The behavior for a plastic material, curve B, is similar to that for many metallic materials; the initial deformation is elastic, which is followed by yielding and a region of plastic deformation. Finally, the deformation displayed by curve C is totally elastic; this rubber-like elasticity (large recoverable strains produced at low stress levels) is displayed by a class of polymers termed the **elastomers.**

Modulus of elasticity (termed *tensile modulus* or sometimes just *modulus* for polymers) and ductility in percent elongation are determined for polymers in the same manner as for metals. For plastic polymers (curve B, Figure 3.9), the yield point is taken as a maximum on the curve, which occurs just beyond the termination of the linear-elastic region (Figure 3.10). The stress at this maximum is the yield strength. Furthermore, tensile strength (TS) corresponds to the stress at which fracture occurs (Figure 3.10); TS may be greater than or less than s_y. Strength, for these plastic polymers, is normally taken as tensile strength. Table 3.4 gives these mechanical properties for several polymeric materials.

Polymers are, in many respects, mechanically dissimilar to metals. For example, the modulus for highly elastic polymeric materials may be as low as 7 MPa but may run as high as 4 GPa for some of the very stiff polymers; modulus values for metals are much larger and range between 48 and 410 GPa. Maximum tensile strengths for polymers are about 100 MPa (15,000 psi) —for some metal alloys 4100 MPa (600,000 psi). And, whereas metals rarely elongate plastically to more than 100%, some highly elastic polymers may experience elongations to greater than 1000%.

Table 3.4 Room-Temperature Mechanical Characteristics of Some of the More Common Polymers

Polymer	Tensile Modulus /GPa(ksi)	Tensile Strength /MPa(ksi)	Yield Strength /MPa(ksi)	Elongation at Break/%
PE(low density)	0.17~0.28(25~41)	8.3~31.4(1.2~4.55)	9.0~14.5(1.3~2.1)	100~650
PE(high density)	1.06~1.09(155~158)	22.1~31.0(3.2~4.5)	26.2~33.1(3.8~4.8)	10~1200
PVC	2.4~4.1(350~600)	40.7~51.7(5.9~7.5)	40.7~44.8(5.9~6.5)	40~80
PTFE	0.40~0.55(58~80)	20.7~34.5(3.0~5.0)	—	200~400
PP	1.14~1.55 (165~225)	31~41.4(4.5~6.0)	31.0~37.2(4.5~5.4)	100~600
PS	2.28~3.28(330~475)	35.9~51.7(5.2~7.5)	—	1.2~2.5
PMMA	2.24~3.24(325~470)	48.3~72.4(7.0~10.5)	53.8~73.1(7.8~10.6)	2.0~5.5
Nylon 6,6	1.58~3.80(230~550)	75.9~94.5(11.0~13.7)	44.8~82.8(6.5~12)	15~300
PET	2.8~4.1(400~600)	48.3~72.4(7.0~10.5)	59.3(8.6)	30~300
PC	2.38(345)	62.8~72.4(9.1~10.5)	62.1(90)	110~150

In addition, the mechanical characteristics of polymers are much more sensitive to temperature changes near room temperature. Consider the stress—strain behavior for poly (methyl methacrylate) (Plexiglas) at several temperatures between 4℃ and 60℃. It should be noted that increasing the temperature produces①a decrease in elastic modulus, ②a reduction in tensile strength, and③an enhancement of ductility—at the material is totally brittle, while there is considerable plastic deformation at both 50℃ and 60℃ (122℉ and 140℉). The influence of strain rate on the mechanical behavior may also be important.

In general, decreasing the rate of deformation has the same influence on the stress—strain characteristics as increasing the temperature; that is, the material becomes softer and more ductile.

3.3.2 Macroscopic Deformation

Many semicrystalline polymers have the spherulitic structure and deform in the following steps:
- elongation of amorphous tie chains
- tilting of lamellar chain folds towards the tensile direction
- separation of crystalline block segments
- orientation of segments and tie chains in the tensile direction

The macroscopic deformation involves an upper and lower yield point and necking. Unlike the case of metals, the neck gets stronger since the deformation aligns the chains so increasing the tensile stress leads to the growth of the neck.

As other mechanical properties, the fracture strength of polymers is much lower than that of metals. Fracture also starts with cracks at flaws, scratches, etc. Fracture involves breaking of covalent bonds in the chains. Thermoplasts can have both brittle and ductile fracture behaviors. Glassy thermosets have brittle fracture at low temperatures and ductile fracture at high temperatures.

Glassy thermoplasts often suffer *grazing* before brittle fracture. Crazes are associated with regions of highly localized yielding which leads to the formation of interconnected microvoids. Crazing absorbs energy thus increasing the fracture strength of the polymer.

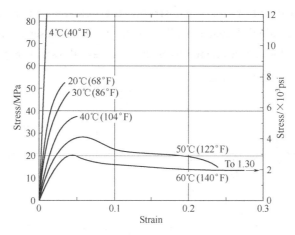

Figure 3.11 The influence of temperature on the stress-strain characteristics of poly (methyl methacrylate). (Adapted from William D. Callister Jr., *Fundamentals of Materials Science and Engineering*, 7th Edition, John Wiley & Sons, Inc., 2007, p.526.)

The tensile stress-strain curve for a semicrystalline material, which was initially undeformed, is shown in Figure 3.11; also included in the figure are schematic representations of the specimen profiles at various stages of deformation. Both upper and lower yield points are evident on the curve, which are followed by a near horizontal region. At the upper yield point, a small neck forms within the gauge section of the specimen. Within this neck, the chains become oriented (i.e., chain axes become aligned parallel to the elongation direction, a condition that is represented schematically in Figure 3.12), which leads to localized strengthening. Consequently, there is a resistance to continued deformation at this point, and specimen elongation proceeds by the propagation of this neck region along the gauge length; the chain orientation phenomenon (Figure 3.12) accompanies this neck extension. This tensile behavior may be contrasted to that found for ductile metals, wherein once a neck has formed, all subsequent deformation is confined to within the neck region.

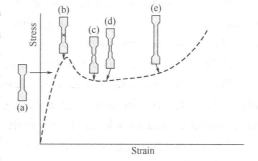

Figure 3.12 Schematic tensile stress-strain curve for a semicrystalline polymer. Specimen contours at several stages of deformation are included. (Adapted from William D. Callister Jr., *Fundamentals of Materials Science and Engineering*, 7th Edition, John Wiley & Sons, Inc., 2007, p.528.)

3.3.3 Viscoelastic Deformation

The word "viscoelastic" is the combination of "viscous" and "elastic". At low temperatures, amorphous polymers deform elastically, like glass, at small elongation. At high temperatures the behavior is viscous, like liquids. At intermediate temperatures, the behavior, like a rubbery solid, is termed viscoelastic.

Elastic deformation is instantaneous, which means that total deformation (or strain) occurs the instant the stress is applied or released (i.e., the strain is independent of time). In addition, upon release of the external stress, the deformation is totally recovered—the specimen assumes its original dimensions. This behavior is represented in Figure 3.13 (b) as strain versus time for the instantaneous load-time curve, shown in Figure 3.13 (a).

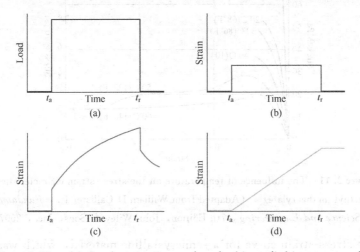

Figure 3.13 (a) Load versus time, where load is applied instantaneously at time and released at For the load—time cycle in (a), the strain-versustime responses are for totally elastic (b), viscoelastic (c), and viscous (d) behaviors.

By way of contrast, for totally viscous behavior, deformation or strain is not instantaneous; that is, in response to an applied stress, deformation is delayed or dependent on time. Also, this deformation is not reversible or completely recovered after the stress is released. This phenomenon is demonstrated in Figure 3.13 (d).

For the intermediate viscoelastic behavior, the imposition of a stress in the manner of Figure 3.13 (a) results in an instantaneous elastic strain, which is followed by a viscous, time-dependent strain, a form of anelasticity; this behavior is illustrated in Figure 3.13 (c).

The reason that polymers are viscoelastic lies in that the movement of a polymer chain is accomplished by successive motion of all the chain segments, which may take a much longer time than the movement of small molecules in metals and ceramics.

A daily-life example of viscoelastic deformation is a phenomenon termed "creep" found for tires and plastic ropes, indicating gradual deformation when the stress level is maintained constant.

An extreme example of these viscoelastic behaviors is found in a silicone polymer that is sold as a novelty and known by some as "silly putty." When rolled into a ball and dropped onto a horizontal surface, it bounces elastically—the rate of deformation during the bounce is very rapid. On the other hand, if pulled in tension with a gradually increasing applied stress, the material elongates or flows like a highly viscous liquid. For this and other viscoelastic materials, the rate of strain determines whether the deformation is elastic or viscous.

3.4 Polymer Types

3.4.1 Plastics

The word plastic is derived from the Greek plastikos meaning capable of being shaped or molded, from plastos meaning molded. It refers to their malleability, or plasticity during manufacture, that allows them to be cast, pressed, or extruded into a variety of shapes—such as filaments, plates, tubes, bottles, boxes, and much more.

Possibly the largest number of different polymeric materials come under the plastic classification. Polyethylene, polypropylene, polyvinyl chloride, polystyrene, and the fluorocarbons, epoxies, phenolics, and polyesters may all be classified as **plastics.** They have a wide variety of combinations of properties. Some plastics are very rigid and brittle; others are flexible, exhibiting both elastic and plastic deformations when stressed, and sometimes experiencing considerable deformation before fracture.

Polymers falling within this classification may have any degree of crystallinity, and all molecular structures and configurations (linear, branched, isotactic, etc.) are possible. Plastic materials may be either thermoplastic or thermosetting; in fact, this is the manner in which they are usually subclassified. The trade names, characteristics, and typical applications for a number of plastics are given as below.

Common plastics and uses are listed as follows:
- Polyester (PES) - Fibers, textiles.
- Polyethylene terephthalate (PET) - Carbonated drinks bottles, peanut butter jars, plastic film, microwavable packaging.
- Polyethylene (PE) - Wide range of inexpensive uses including supermarket bags, plastic bottles.
- High-density polyethylene - Detergent bottles and milk jugs.
- Polyvinyl chloride (PVC) - Plumbing pipes and guttering, shower curtains, window frames, flooring.
- Polyvinylidene chloride (PVDC) (Saran) - Food packaging.
- Low-density polyethylene (LDPE) - Outdoor furniture, siding, floor tiles, shower curtains, clamshell packaging.
- Polypropylene (PP) - Bottle caps, drinking straws, yogurt containers, appliances, car fenders (bumpers), plastic pressure pipe systems.
- Polystyrene (PS) - Packaging foam/ "peanuts", food containers, plastic tableware, disposable cups, plates, cutlery, CD and cassette boxes.
- High impact polystyrene (HIPS) -: Refrigerator liners, food packaging, vending cups.

- Polyamides (PA) (Nylons) - Fibers, toothbrush bristles, fishing line, under-the-hood car engine moldings.
- Acrylonitrile butadiene styrene (ABS) - Electronic equipment cases (e.g., computer monitors, printers, keyboards), drainage pipe.
- Polycarbonate (PC) - Compact discs, eyeglasses, riot shields, security windows, traffic lights, lenses.
- Polycarbonate/Acrylonitrile Butadiene Styrene (PC/ABS) - A blend of PC and ABS that creates a stronger plastic. Used in car interior and exterior parts, and mobile phone bodies.
- Polyurethanes (PU) - Cushioning foams, thermal insulation foams, surface coatings, printing rollers (Currently 6th or 7th most commonly used plastic material, for instance the most commonly used plastic found in cars).

Several plastics exhibit especially outstanding properties. For applications in which optical transparency is critical, polystyrene and polymethyl methacrylate are especially well suited; however, it is imperative that the material be highly amorphous or, if semicrystalline, have very small crystallites. The fluorocarbons have a low coefficient of friction and are extremely resistant to attack by a host of chemicals, even at relatively high temperatures. They are utilized as coatings on nonstick cookware, in bearings and bushings, and for high-temperature electronic components.

3.4.2 Elastomers

An elastomer is a polymer that can be deformed to very large strains and the spring back elastically to the original length, a behavior first observed in natural rubber. Elastic elongation is due to uncoiling, untwisting and straightening of chains in the stress direction.

To be elastomeric, the polymer needs to meet several criteria:
- must not crystallize easily
- have relatively free chain rotations
- delayed plastic deformation by cross-linking (achieved by vulcanization).
- be above the glass transition temperature

Pproperties and applications of common elastomers are typical and, of course, depend on the degree of vulcanization and on whether any reinforcement is used. Natural rubber is still utilized to a large degree because it has an outstanding combination of desirable properties. However, the bile tires, reinforced with carbon black, NBR, which is highly resistant to degradation and swelling, is another common synthetic elastomer.

For many applications (e.g., automobile tires), the mechanical properties of even vulcanized rubbers are not satisfactory in terms of tensile strength, abrasion and tear resistance, and stiffness. These characteristics may be further improved by additives such as carbon black.

Finally, some mention should be made of the silicone rubbers such as PDMS. Of course, as elastomers, these materials are crosslinked. The silicone elastomers possess a high degree of flexibility at low temperatures [to $-90^\circ C$ ($-130^\circ F$)] and yet are stable to temperatures as high as 250℃ (480°F). In addition, they are resistant to weathering and lubricating oils. A further attractive characteristic is that some silicone rubbers vulcanize at room temperature (RTV rubbers).

3.4.3 Fibers

The **fiber** polymers are capable of being drawn into long filaments having at least a 100∶1 length-to-diameter ratio. Most commercial fiber polymers are utilized in the textile industry, being woven or knit into cloth or fabric. In addition, the aramid fibers are employed in composite materials. To be useful as a textile material, a fiber polymer must have a host of rather restrictive physical and chemical properties. While in use, fibers may be subjected to a variety of mechanical deformations—stretching, twisting, shearing, and abrasion. Consequently, they must have a high tensile strength (over a relatively wide temperature range) and a high modulus of elasticity, as well as abrasion resistance. These properties are governed by the chemistry of the polymer chains and also by the fiber drawing process.

The molecular weight of fiber materials should be relatively high. Also, since the tensile strength increases with degree of crystallinity, the structure and configuration of the chains should allow the production of a highly crystalline polymer; that translates into a requirement for linear and unbranched chains that are symmetrical and have regularly repeating mer units.

Convenience in washing and maintaining clothing depends primarily on the thermal properties of the fiber polymer, that is, its melting and glass transition temperatures. Furthermore, fiber polymers must exhibit chemical stability to a rather extensive variety of environments, including acids, bases, bleaches, dry cleaning solvents, and sunlight. In addition, they must be relatively nonflammable and amenable to drying.

3.4.4 Miscellaneous Applications Coatings

Coating is a covering that is applied to the surface of an object, usually referred to as the substrate. Coatings are frequently applied to improve surface properties of the substrate, such as appearance, adhesion, wetability, corrosion resistance, wear resistance, and scratch resistance. In other cases, in particular in printing processes and semiconductor device fabrication (where the substrate is a wafer), the coating forms an essential part of the finished product.

Many of the ingredients in coating materials are polymers, the majority of which are organic in origin. These organic coatings fall into several different classifications, as follows: paint, varnish, enamel, lacquer, and shellac.

Adhesives

An **adhesive** is a substance used to join together the surfaces of two solid materials (termed "adherends") to produce a joint with a high shear strength. The bonding forces between the adhesive and adherend surfaces are thought to be electrostatic, similar to the secondary bonding forces between the molecular chains in thermoplastic polymers. Even though the inherent strength of the adhesive may be much less than that of the adherend materials, nevertheless, a strong joint may be produced if the adhesive layer is thin and continuous. If a good joint is formed, the adherend material may fracture or rupture before the adhesive.

Polymeric materials that fall within the classifications of thermoplastics, thermosetting resins, elastomeric compounds, and natural adhesives (animal glue, casein, starch, and rosin) may serve adhesive functions. Polymer adhesives may be used to join a large variety of material combinations: metal—metal, metal—plastic, metal—ceramic, and so on. The primary drawback is the service temperature limitation. Organic polymers maintain their mechanical integrity only at relatively low temperatures, and strength decreases rapidly with increasing temperature.

Films

Within relatively recent times, polymeric materials have found widespread use in the form of thin *films*. Films having thicknesses between 0.025 and 0.125 mm (0.001 and 0.005 in.) are fabricated and used extensively as bags for packaging food products and other merchandise, as textile products, and a host of other uses. Important characteristics of the materials produced and used as films include low density, a high degree of flexibility, high tensile and tear strengths, resistance to attack by moisture and other chemicals, and low permeability to some gases, especially water vapor. Some of the polymers that meet these criteria and are manufactured in film form are polyethylene, polypropylene, cellophane, and cellulose acetate.

Foams

Foams are plastic materials that contain a relatively high volume percent of small pores. Both thermoplastic and thermosetting materials are used as foams; these include polyurethane, rubber, polystyrene, and polyvinyl chloride. Foams are commonly used as cushions in automobiles and furniture as well as in packaging and thermal insulation. The foaming process is carried out by incorporating into the batch of material a blowing agent that upon heating, decomposes with the liberation of a gas. Gas bubbles are generated throughout the now-fluid mass, which remain as pores upon cooling and give rise to a spongelike structure. The same effect is produced by bubbling an inert gas through a material while it is in a molten state.

3.5 Processing of Polymers

The large macromolecules of the commercially useful polymers must be synthesized from substances having smaller molecules in a process termed polymerization. Furthermore, the properties of a polymer may be modified and enhanced by the inclusion of additive materials. Finally, a finished piece having a desired shape must be fashioned during a forming operation. This section treats polymerization processes and the various forms of additives, as well as specific forming procedures.

3.5.1 Polymerization

Polymerization is a process of reacting monomer molecules together in a chemical reaction to form three-dimensional networks or polymer chains. Most generally, the raw materials for synthetic polymers are derived from coal and petroleum products, which are composed of molecules having low molecular weights. The reactions by which polymerization occurs are grouped into two general classifications—addition and condensation—according to the reaction mechanism, as discussed below.

Addition Polymerization

Addition polymerization (sometimes called *chain reaction polymerization*) is a process by which bifunctional monomer units are attached one at a time in chainlike fashion to form a linear macromolecule; the composition of the resultant product molecule is an exact multiple for that of the original reactant monomer.

Three distinct stages—initiation, propagation, and termination—are involved in addition polymerization. During the initiation step, an active center capable of propagation is formed by a reaction between an initiator (or catalyst) species and the monomer unit. This process has already been demonstrated for polyethylene [Equation (3.5)], which is repeated as follows:

$$R\cdot + \begin{array}{c} H \\ | \\ C \\ | \\ H \end{array}=\begin{array}{c} H \\ | \\ C \\ | \\ H \end{array} \longrightarrow R-\begin{array}{c} H \\ | \\ C \\ | \\ H \end{array}-\begin{array}{c} H \\ | \\ C\cdot \\ | \\ H \end{array} \qquad (3.5)$$

Again, R· represents the active initiator, and · is an unpaired electron.

Propagation involves the linear growth of the molecule as monomer units become attached to one another in succession to produce the chain molecule, which is represented, again for polyethylene, as follows:

$$R-\begin{array}{c}H\\|\\C\\|\\H\end{array}-\begin{array}{c}H\\|\\C\cdot\\|\\H\end{array}+\begin{array}{c}H\\|\\C\\|\\H\end{array}=\begin{array}{c}H\\|\\C\\|\\H\end{array} \longrightarrow R-\begin{array}{c}H\\|\\C\\|\\H\end{array}-\begin{array}{c}H\\|\\C\\|\\H\end{array}-\begin{array}{c}H\\|\\C\\|\\H\end{array}-\begin{array}{c}H\\|\\C\cdot\\|\\H\end{array} \qquad (3.6)$$

Chain growth is relatively rapid; the period required to grow a molecule consisting of, say, 1000 mer units is on the order of 10^{-2} to 10^{-3} s.

Propagation may end or terminate in different ways. First, the active ends of two propagating chains may react or link together to form a nonreactive molecule, as follows:

$$R-C-C-C-C\cdot + \cdot C-C-C-C-R \longrightarrow$$
$$R-C-C-C-C-C-C-C-C-R \qquad (3.7)$$

thus terminating the growth of each chain. Or, an active chain end may react with an initiator or other chemical species having a single active bond, as follows:

$$-C-C-C-C\cdot + \cdot R \longrightarrow -C-C-C-C-R \qquad (3.8)$$

with the resultant cessation of chain growth.

Molecular weight is governed by the relative rates of initiation, propagation, and termination. Ordinarily, they are controlled to ensure the production of a polymer having the desired degree of polymerization. Addition polymerization is used in the synthesis of polyethylene, polypropylene, polyvinyl chloride, and polystyrene, as well as many of the copolymers.

Condensation Polymerization

Condensation (or *step reaction*) **polymerization** is the formation of polymers by stepwise intermolecular chemical reactions that normally involve more than one monomer species; there is usually a small molecular weight by-product such as water, which is eliminated. No reactant species has the chemical formula of the mer repeat unit, and the intermolecular reaction occurs every time a mer repeat unit is formed. For example, consider the formation of a polyester, poly (ethylene terephthalate) (PET), from the reaction between ethylene glycol and terephthalic acid; the intermolecular reaction is as follows:

$$\text{ethylene glycol} \quad \text{tetephthalic acid}$$

$$\text{HO}-\underset{\underset{H}{|}}{\overset{\overset{H}{|}}{C}}-\underset{\underset{H}{|}}{\overset{\overset{H}{|}}{C}}-\text{OH} \quad \text{HO}-\overset{O}{\overset{\|}{C}}-\underset{}{\bigcirc}-\overset{O}{\overset{\|}{C}}-\text{OH} \tag{3.9}$$

$$\longrightarrow \text{HO}-\underset{\underset{H}{|}}{\overset{\overset{H}{|}}{C}}-\underset{\underset{H}{|}}{\overset{\overset{H}{|}}{C}}-\text{O}-\overset{O}{\overset{\|}{C}}-\underset{}{\bigcirc}-\overset{O}{\overset{\|}{C}}-\text{OH} + \text{H}_2\text{O}$$

This stepwise process is successively repeated, producing, in this case, a linear molecule. The chemistry of the specific reaction is not important, but rather, the condensation polymerization mechanism. Furthermore, reaction times for condensation are generally longer than for addition polymerization.

Condensation reactions often produce trifunctional monomers capable of forming crosslinked and network polymers. The thermosetting polyesters and phenolformaldehyde, the nylons, and the polycarbonates are produced by condensation polymerization. Some polymers, such as nylon, may be polymerized by either technique.

3.5.2 Polymer Additives

Most of the properties of polymers discussed previously are intrinsic ones—that is, characteristic of or fundamental to the specific polymer. Some of these properties are related to and controlled by the molecular structure. Many times, however, it is necessary to modify the mechanical, chemical, and physical properties to a much greater degree than is possible by the simple alteration of this fundamental molecular structure. Foreign substances called *additives* are intentionally introduced to enhance or modify many of these properties, and thus render a polymer more serviceable. Typical additives include filler materials, plasticizers, stabilizers, colorants, and flame retardants.

Fillers

Filler materials are most often added to polymers to improve tensile and compressive strengths, abrasion resistance, toughness, dimensional and thermal stability, and other properties. Materials used as particulate fillers include wood flour (finely powdered sawdust), silica flour and sand, glass, clay, talc, limestone, and even some synthetic polymers. Particle sizes range all the way from 10nm to macroscopic dimensions. Because these inexpensive materials replace some volume of the more expensive polymer, the cost of the final product is reduced.

Plasticizers

The flexibility, ductility, and toughness of polymers may be improved with the aid of additives called **plasticizers**. Their presence also produces reductions in hardness and stiffness. Plasticizers are generally liquids having low vapor pressures and low molecular weights. The small plasticizer molecules occupy positions between the large polymer chains, effectively increasing the interchain distance with a reduction in the secondary intermolecular bonding. Plasticizers are commonly used in polymers that are intrinsically brittle at room temperature, such as polyvinyl chloride and some of the acetate copolymers. In effect, the plasticizer lowers the glass transition temperature, so that at ambient conditions the polymers may be used in applications requiring some degree of pliability and ductility. These applications include thin sheets or films, tubing, raincoats, and curtains.

Stabilizers

Some polymeric materials, under normal environmental conditions, are subject to rapid deterioration, generally in terms of mechanical integrity. Most often, this deterioration is a result of exposure to light, in particular ultraviolet radiation, and also oxidation. Ultraviolet radiation interacts with, and causes a severance of some of the covalent bonds along the molecular chain, which may also result in some crosslinking. Oxidation deterioration is a consequence of the chemical interaction between oxygen atoms and the polymer molecules. Additives that counteract these deteriorative processes are called **stabilizers**.

Colorants

Colorants impart a specific color to a polymer; they may be added in the form of dyes or pigments. The molecules in a dye actually dissolve and become part of the molecular structure of the polymer. Pigments are filler materials that do not dissolve, but remain as a separate phase; normally they have a small particle size, are transparent, and have a refractive index near to that of the parent polymer. Others may impart opacity as well as color to the polymer.

Flame Retardants

The flammability of polymeric materials is a major concern, especially in the manufacture of textiles and children's toys. Most polymers are flammable in their pure form; exceptions include those containing significant contents of chlorine and/or fluorine, such as polyvinyl chloride and polytetrafluoroethylene. The flammability resistance of the remaining combustible polymers may be enhanced by additives called **flame retardants**. These retardants may function by interfering with the combustion process through the gas phase, or by initiating a chemical reaction that causes a cooling of the combustion region and a cessation of burning.

3.5.3 Forming Techniques for Plastics

Quite a variety of different techniques are employed in the forming of polymeric materials. The method used for a specific polymer depends on several factors: ①whether the material is thermoplastic or thermosetting; ②if thermoplastic, the temperature at which it softens; ③the atmospheric stability of the material being formed; and④the geometry and size of the finished product. There are numerous similarities between some of these techniques and those utilized for fabricating metals and ceramics.

Fabrication of polymeric materials normally occurs at elevated temperatures and often by the application of pressure. Thermoplastics are formed above their glass transition temperatures, if amorphous, or above their melting temperatures, if semicrystalline; an applied pressure must be maintained as the piece is cooled so that the formed article will retain its shape. One significant economic benefit of using thermoplastics is that they may be recycled; scrap thermoplastic pieces may be re-melted and reformed into new shapes.

Fabrication of thermosetting polymers is ordinarily accomplished in two stages. First comes the preparation of a linear polymer (sometimes called a prepolymer) as a liquid, having a low molecular weight. This material is converted into the final hard and stiff product during the second stage, which is normally carried out in a mold having the desired shape. This second stage, termed "curing," may occur during heating and/or by the addition of catalysts, and often under pressure. During curing, chemical and structural changes occur on a molecular level: a crosslinked or a network structure forms. After curing, thermoset polymers may be removed from a mold while still hot, since they are now dimensionally stable. Thermosets are difficult to recycle, do not melt, are usable at higher temperatures than thermoplastics, and are more chemically inert.

Molding is the most common method for forming plastic polymers. The several molding techniques used include compression, transfer, blow, injection, and extrusion molding. For each, a finely pelletized or granulized plastic is forced, at an elevated temperature and by pressure, to flow into, fill, and assume the shape of a mold cavity.

Compression and Transfer Molding

Compression and Transfer Molding is accomplished by placing a pre-weighed amount of rubber in a matched metal mold and closing the mold. The heat and pressure cause the rubber to liquefy and flow into the voids in the tool where it chemically reacts and hardens into the final shape. Very large shapes can be molded in compression presses.

Compression molding is a method of molding in which the molding material, generally preheated, is first placed in an open, heated mold cavity, as illustrated in Figure 3.14. The mold is closed with a top force or plug member, pressure is applied to force the material into contact with all mold areas, while heat and pressure are maintained until the molding material has cured. The process employs thermosetting resins in a partially cured stage, either in the form of granules, putty-like masses, or preforms.

For compression molding, the appropriate amounts of thoroughly mixed polymer and necessary additives are placed between male and female mold members. Both mold pieces are heated; however, only one is movable. The mold is closed, and heat and pressure are applied, causing the plastic material to become viscous and conform to the mold shape. Before molding, raw materials may be mixed and cold pressed into a disc, which is called a preform. Preheating of the preform reduces molding time and pressure, extends the die lifetime, and produces a more uniform finished piece. This molding technique lends itself to the fabrication of both thermoplastic and thermosetting polymers; however, its use with thermoplastics is more time consuming and expensive.

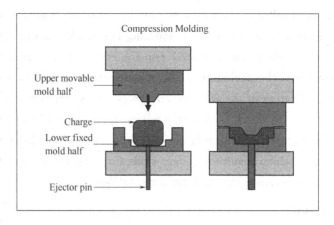

Figure 3.14 Schematic diagram of a compression molding apparatus

In transfer molding, a variation of compression molding, the solid ingredients are first melted in a heated transfer chamber. As the molten material is injected into the mold chamber, the pressure is distributed more uniformly over all surfaces. This process is used with thermosetting polymers and for pieces having complex geometries.

Injection Molding

Injection molding is a manufacturing process for producing parts from both thermoplastic and thermosetting plastic materials. Material is fed into a heated barrel, mixed, and forced into a mold cavity where it cools and hardens to the configuration of the mold cavity.

After a product is designed, usually by an industrial designer or an engineer, molds are made by a moldmaker (or toolmaker) from metal, usually either steel or aluminum, and precision-machined to form the features of the desired part. Injection molding is widely used for manufacturing a variety of parts, from the smallest component to entire body panels of cars. A schematic cross section of the apparatus used is illustrated in Figure 3.15.

The correct amount of pelletized material is fed from a loading hopper into a cylinder by the motion of a plunger or ram. This charge is pushed forward into a heating chamber, at which point the thermoplastic material melts to form a viscous liquid. Next, the molten plas-

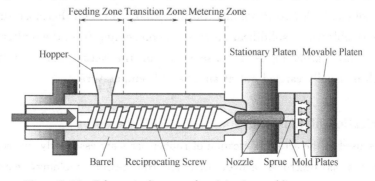

Figure 3.15 Schematic diagram of an injection molding apparatus

tic is impelled, again by ram motion, through a nozzle into the enclosed mold cavity; pressure is maintained until the molding has solidified. Finally, the mold is opened, the piece is ejected, the mold is closed, and the entire cycle is repeated. Probably the most outstanding feature of this technique is the speed with which pieces may be produced. For thermoplastics, solidification of the injected charge is almost immediate; consequently, cycle times for this process are short (commonly within the range of 10 to 30 s). Thermosetting polymers may also be injection molded; curing takes place while the material is under pressure in a heated mold, which results in longer cycle times than for thermoplastics. This process is sometimes termed reaction injection molding (RIM).

Extrusion

Plastics extrusion is a high volume manufacturing process in which raw plastic material is melted and formed into a continuous profile. Extrusion produces items such as pipe/tubing, weather stripping, fence, deck railing, window frames, adhesive tape and wire insulation. The extrusion process is simply injection molding of a viscous thermoplastic through an open-ended die, similar to the extrusion of metals. A mechanical screw or auger propels through a chamber the pelletized material, which is successively compacted, melted, and formed into a continuous charge of viscous fluid. Extrusion takes place as this molten mass is forced through a die orifice. Solidification of the extruded length is expedited by blowers or a water spray just before passing onto a moving conveyor. The technique is especially adapted to producing continuous lengths having constant cross-sectional geometries, for example, rods, tubes, hose channels, sheets, and filaments.

Blow Molding

The blow molding process for the fabrication of plastic containers is similar to that used for blowing glass bottles. First, a parison, or length of polymer tubing is extruded. While still in a semimolten state, the parison is placed in a two-piece mold having the desired container configuration. The hollow piece is formed by blowing air or steam under pressure into the parison, forcing the tube walls to conform to the contours of the mold. Of course the temperature and viscosity of the parison must be carefully regulated.

Casting

Like metals, polymeric materials may be cast, as when a molten plastic material is poured into a mold and allowed to solidify. Both thermoplastic and thermosetting plastics may be cast. For thermoplastics, solidification occurs upon cooling from the molten state; however, for thermosets, hardening is a consequence of the actual polymerization or curing process, which is usually carried out at an elevated temperature.

3.5.4 Fabrication of Elastomers

Techniques used in the actual fabrication of rubber parts are essentially the same as those discussed for plastics as described above, that is, compression molding, extrusion, and so on. Furthermore, most rubber materials are vulcanized and some are reinforced with carbon black.

3.5.5 Fabrication of Fibers and Films

Fibers

The process by which fibers are formed from bulk polymer material is termed **spinning**. Most often, fibers are spun from the molten state in a process called melt spinning. The material to be spun is first heated until it forms a relatively viscous liquid. Next, it is pumped down through a plate called a spinnerette, which contains numerous small, round holes. As the molten material passes through each of these orifices, a single fiber is formed, which solidifies almost immediately upon passing into the air.

The crystallinity of a spun fiber will depend on its rate of cooling during spinning. The strength of fibers is improved by a postforming process called **drawing**. Again, drawing is simply the mechanical elongation of a fiber in the direction of its axis. During this process the molecular chains become oriented in the direction of drawing, such that the tensile strength, modulus of elasticity, and toughness are improved. Although the mechanical strength of a drawn fiber is improved in this axial direction, strength is reduced in a transverse or radial direction. However, since fibers are normally stressed only along the axis, this strength differential is not critical. The cross section of drawn fibers is nearly circular, and the properties are uniform throughout the cross section.

Films

Many films are simply extruded through a thin die slit; this may be followed by a rolling operation that serves to reduce thickness and improve strength. Alternatively, film may be blown: continuous tubing is extruded through an annular die; then, by maintaining a carefully controlled positive gas pressure inside the tube, wall thickness may be continuously reduced to produce a thin cylindrical film, which may be cut and laid flat. Some of the newer films are produced by coextrusion; that is, multilayers of more than one polymer type are extruded simultaneously.

Summary

Hydrocarbon Molecules

Polymer Molecules

The Chemistry of Polymer Molecules

Most polymeric materials are composed of very large molecules—chains of carbon atoms, to which are side-bonded various atoms or radicals. These macromolecules may be thought of as being composed of mers, smaller structural entities, which are repeated along the chain. Mer structures of some of the chemically simple polymers (e.g., polyethylene, polytetrafluoroethylene, polyvinyl chloride, and polypropylene) were presented.

Molecular Weight

Molecular weights for high polymers may be in excess of a million. Since all molecules are not of the same size, there is a distribution of molecular weights. Molecular weight is often expressed in terms of number and weight averages. Chain length may also be specified by degree of polymerization, the number of mer units per average molecule.

Molecular Shape
Molecular Structure
Copolymers

Several molecular characteristics that have an influence on the properties of polymers were discussed. Molecular entanglements occur when the chains assume twisted, coiled, and kinked shapes or contours. With regard to molecular structure, linear, branched, crosslinked, and network structures are possible. The copolymers include random, alternating, block, and graft types.

Thermoplastic and Thermosetting Polymers

With regard to behavior at elevated temperatures, polymers are classified as either thermoplastic or thermosetting. The former have linear and branched structures; they soften when heated and harden when cooled. In contrast, thermosets, once having hardened, will not soften upon heating; their structures are crosslinked and network.

Polymer Crystallinity

When the packing of molecular chains is such as to produce an ordered atomic arrangement, the condition of crystallinity is said to exist. In addition to being entirely amorphous, polymers may also exhibit varying degrees of crystallinity; for the latter case, crystalline regions are interdispersed within amorphous areas. Crystallinity is facilitated for polymers that are chemically simple and that have regular and symmetrical chain structures.

Stress-Strain Behavior

On the basis of stress-strain behavior, polymers fall within three general classifications: brittle, plastic, and highly elastic. These materials are neither as strong nor as stiff as metals, and their mechanical properties are sensitive to changes in temperature and strain rate. However, their high flexibilities, low densities, and resistance to corrosion make them the materials of choice for many applications.

Viscoelastic Deformation

Viscoelastic mechanical behavior, being intermediate between totally elastic and totally viscous, is displayed by a number of polymeric materials. It is characterized by the relaxation modulus, a time-dependent modulus of elasticity. The magnitude of the relaxation modulus is very sensitive to temperature; critical to the in-service temperature range for elastomers is this temperature dependence.

Crystallization
Melting
The Glass Transition
Melting and Glass Transition Temperatures

The molecular mechanics of crystallization, melting, and the glass transition were discussed. The manner in which melting and glass transition temperatures are determined was outlined; these parameters are important relative to the temperature range over which a particular polymer may be utilized and processed. The magnitudes of T_m and T_g increase with increasing chain stiffness; stiffness is enhanced by the presence of chain double bonds and side groups that are either bulky or polar. Molecular weight and degree of branching also affect T_m and T_g.

Plastics

The various types and applications of polymeric materials were also discussed. Plastic materials are perhaps the most widely used group of polymers, which include the following: polyethylene, polypropylene, poly (vinyl chloride), polystyrene, and the fluorocarbons, epoxies, phenolics, and polyesters.

Fibers

Many polymeric materials may be spun into fibers, which are used primarily in textiles. Mechanical, thermal, and chemical characteristics of these materials are especially critical.

Miscellaneous Applications

Other miscellaneous applications that employ polymers include coatings, adhesives, films, and foams.

Polymerization

Polymer Additives

The final sections of this chapter treated synthesis and fabrication techniques for polymeric materials. Synthesis of large molecular weight polymers is attained by polymerization, of which there are two types: addition and condensation. The properties of polymers may be further modified by using additives; these include fillers, plasticizers, stabilizers, colorants, and flame retardants.

Forming Techniques for Plastics

Fabrication of plastic polymers is usually accomplished by shaping the material in molten form at an elevated temperature, using at least one of several different molding techniques—compression, transfer, injection, and blow. Extrusion and casting are also possible.

Fabrication of Fibers and Films

Some fibers are spun from a viscous melt, after which they are plastically elongated during a drawing operation, which improves the mechanical strength. Films are formed by extrusion and blowing, or by calendering.

Important Terms and Concepts

Alternating Copolymer	Graft Copolymer	Polymer Crystallinity
Bifunctional	Homopolymer	Random Copolymer
Trifunctional	Isomerism	Repeat Unit
Block Copolymer	Isotactic Configuration	Saturated
Branched Polymer	Linear Polymer	Spherulite
Chain-Folded Model	Macromolecule	Stereoisomerism
Unsaturated	Molecular Chemistry	Syndiotactic Configuration
Copolymer	Molecular Structure	Thermoplastic Polymer
Crosslinked Polymer	Molecular Weight	Thermosetting Polymer
Crystallite	Monomer	Trans(Structure)
Degree of Polymerization	Network Polymer	Functionality
Addition Polymerization	Flame Retardant	Relaxation Modulus
Adhesive	Foam	Spinning
Colorant	Glass Transition Temperature	Stabilizer
Condensation Polymerization	Liquid Crystal Polymer	Thermoplastic Elastomer
Drawing	Melting Temperature	Ultrahigh Molecular Weight
Elastomer	Molding	Polyethylene
Fiber	Plasticizer	Viscoelasticity
Filler	Plastic	

References

[1] Baer, E., "Advanced Polymers," *Scientific American*, Vol. 255, No. 4, October 1986, pp. 178-190.

[2] Carraher, C. E., Jr., *Seymour/Carraher's Polymer Chemistry*, 6th edition, Marcel Dekker, New York, 2003.

[3] Cowie, J. M. G., *Polymers: Chemistry and Physics of Modern Materials*, 2nd edition, Chapman and Hall (USA), New York, 1991. 4. McCrum, N. G., C. P. Buckley, and C. B. Bucknall, *Principles of Polymer Engineering*, 2nd edition, Oxford University Press, Oxford, 1997. Chapters 0-6.

[4] Rodriguez, F., C. Cohen, C. K. Ober, and L. Archer, *Principles of Polymer Systems*, 5th edition, Taylor & Francis, New York, 2003.

[5] Rosen, S. L., *Fundamental Principles of Polymeric Materials*, 2nd edition, Wiley, New York, 1993.

[6] Rudin, A., *The Elements of Polymer Science and Engineering*, 2nd edition, Academic Press, San Diego, 1998.

[7] Sperling, L. H., *Introduction to Physical Polymer Science*, 3rd edition, Wiley, New York, 2001.

[8] Young, R. J., and P. Lovell, *Introduction to Polymers*, 2nd edition, Chapman and Hall, New York, 1991.

[9] Billmeyer, F. W., Jr., *Textbook of Polymer Science*, 3rd edition, Wiley-Interscience, New York, 1984.

[10] Carraher, C. E., Jr., *Seymour/Carraher's Polymer Chemistry*, 6th edition, Marcel Dekker, New York, 2003.

[11] Harper, C. A. (Editor), *Handbook of Plastics, Elastomers and Composites*, 3rd edition, McGraw-Hill Professional Book Group, New York, 1996.

[12] Landel, R. F. (Editor), *Mechanical Properties of Polymers and Composites*, 2nd edition, Marcel Dekker, New York, 1994.

[13] McCrum, N. G., C. P. Buckley, and C. B. Bucknall, *Principles of Polymer Engineering*, 2nd edition, Oxford University Press, Oxford, 1997. Chapters 7-8.

[14] Muccio, E. A., *Plastic Part Technology*, ASM International, Materials Park, OH, 1991.
Muccio, E. A., *Plastics Processing Technology*, ASM International, Materials Park, OH, 1994.

[15] Powell, P. C., and A. J. Housz, *Engineering with Polymers*, 2nd edition, Nelson Thornes, Cheltenham, UK, 1998.

[16] Rosen, S. L., *Fundamental Principles of Polymeric Materials*, 2nd edition, Wiley, New York, 1993.

[17] Rudin, A., *The Elements of Polymer Science and Engineering*, 2nd edition, Academic Press, San Diego, 1998.

[18] Strong, A. B., *Plastics: Materials and Processing*, 3rd edition, Prentice Hall PTR, Paramus, IL, 2006.

[19] Tobolsky, A. V., *Properties and Structures of Polymers*, Wiley, New York, 1960.

[20] Ward, I. M. and J. Sweeney, *An Introduction to the Mechanical Properties of Solid Polymers*, 2nd edition, John Wiley & Sons, Hoboken, NJ, 2004.

[21] *Engineered Materials Handbook*, Vol. 2, *Engineering Plastics*, ASM International, Materials Park, OH, 1988.

[22] William D. Callister Jr., *Fundamentals of Materials Science and Engineering*, 7th Edition, John Wiley & Sons, Inc., 2007.

Chapter 4　Metallic Materials

Learning Objectives
After studying this chapter you should be able to do the following:
1. *Define engineering stress and engineering strain.*
2. *State Hooke's law, and note the conditions under which it is valid.*
3. *Given an engineering stress—strain diagram, determine (a) the modulus of elasticity, (b) the yield strength (0.002 strain offset), and (c) the tensile strength, and (d) estimate the percent elongation.*
4. *Describe edge and screw dislocation motion from an atomic perspective.*
5. *Describe how plastic deformation occurs by the motion of edge and screw dislocations in response to applied shear stresses.*
6. *Describe and explain solid-solution strengthening for substitutional impurity atoms in terms of lattice strain interactions with dislocations.*
7. *Describe and explain the phenomenon of strain hardening (or cold working) in terms of dislocations and strain field interactions.*
8. *Describe the mechanism of crack propagation for both ductile and brittle modes of fracture.*
9. *Define fracture toughness in terms of (a) a brief statement, and (b) an equation; define all parameters in this equation.*
10. *Define fatigue and specify the conditions under which it occurs.*
11. *Define creep and specify the conditions under which it occurs.*
12. *(a) Schematically sketch simple isomorphous and eutectic phase diagrams.*
    (b) On these diagrams label the various phase regions.
    (c) Label liquidus, solidus, and solvus lines.
13. *Name and describe different types of metal alloys, for each, cite compositional differences, distinctive properties, and typical applications.*
14. *Name and describe forming operations that are used to fabricate metal alloys.*

4.1　Mechanical Properties of Metals

4.1.1　Introduction

　　Metallic materials are generally divided into two groups being ferrous and non-ferrous. Ferrous metals include steel and pig iron (with a carbon content of a few percent) and alloys of iron with other metals (such as stainless steel). Non-ferrous metals do not contain iron as a main ingredient, such as aluminum and copper alloys. A simple test for ferrous/non-ferrous materials is to use magnet as a magnet will sick to ferrous materials due to its iron content.

The most common metal, steel has a wide range of strengths and ductilities (mechanical properties), which makes it the material of choice for numerous applications. While low-carbon steel is used as reinforcing bars in concrete and in the body of automobiles, quenched and tempered high-carbon steel is used in more critical applications such as axles and gears. Cast iron, much more brittle, is used in a variety of applications, including automobile engine blocks. These different applications require, obviously, different mechanical properties of the material.

The mechanical behavior of a material reflects the relationship between its response or deformation to an applied load or force. Important mechanical properties are strength, hardness, ductility, and stiffness. Those factors are often called upon to design structures/components using predetermined materials such that unacceptable levels of deformation and/or failure will not occur.

The mechanical properties are of concern to a variety of parties (e. g. , producers and consumers of materials, research organizations, and government agencies) that have differing interests. Consequently, it is imperative that there be some consistency in the manner in which tests are conducted, and in the interpretation of their results. This consistency is accomplished by using standardized testing techniques. In the USA, the American Society for Testing Materials (ASTM) publishes standard specifications and methods of testing which are updated every three years. In the UK, the British Standards Institution (BSI) publishes an annual catalogue of all BSI Standards, and agreed European Standards (EN series).

4.1.2 Tensile Test

A sophisticated tensile tester is routinely used by virtually all industrial and scientific labs. In the tester, a standard sample with rod-shaped or flat piece under investigation is held between a fixed and a movable arm as shown in Figure 4.1. A force upon the test piece is exerted by slowly driving the movable cross-head away from the fixed arm. This causes a **stress**, σ, on the sample, which is defined to be the force, F, per unit area, A_0, that is,

$$\sigma = \frac{F}{A_0} \qquad (4.1)$$

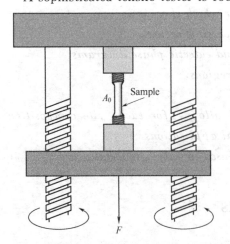

Figure 4.1 Schematic representation of a tensile test equipment. The lower cross-bar is made to move downward and thus extecds a force, F, on the test piece whose cross-sectional area is A_0. The specimen to be tested is either threaded into the specimen holders or held by a vice grip

Since the cross section changes during the tensile test, the initial unit area, A_0, is mostly used. If the force is applied parallel to the axis of a rod-shaped material, as in the tensile tester (that is, perpendicular to the faces A_0), σ is called a **tensile stress**. If the stress is applied parallel to the faces, it is termed **shear stress**, t.

Many materials respond to stress by changing

their dimensions. Under tensile stress, the rod becomes longer in the direction of the applied force (and eventually narrower perpendicular to that axis). The change in longitudinal dimension in response to stress is called strain, e, that is:

$$\varepsilon = \frac{l - l_0}{l_0} = \frac{\Delta l}{l_0} \quad (4.2)$$

where l_0 is the initial length of the rod and l is its final length. The absolute value of the ratio between the lateral strain (shrinkage) and the longitudinal strain (elongation) is called the **Poisson ratio**, ν. Its maximum value is 0.5 (no net volume change). In reality, the Poisson ratio for metals and alloys is generally between 0.27 and 0.35; in plastics (e.g., nylon) it may be as large as 0.4; and for rubbers it is even 0.49, which is near the maximum possible value.

The force is measured in newtons ($1N = 1kg \cdot m/s^2$) and the stress is given in N/m^2 or pascal (Pa). (Engineers in the United States occasionally use the pounds per square inch (psi) instead, where 1 psi $= 6.895 \times 10^3$ Pa and 1 pound $= 4.448N$.) The strain is unitless, as can be seen from Equation 4.2 and is usually given in percent of the original length.

The result of a tensile test is commonly displayed in a *stress-strain diagram* as schematically depicted in Figure 4.2. Several important characteristics are immediately evident. During the initial stress period, the elongation of the material responds to s in a linear fashion; the rod reverts back to its original length upon relief of the load. This region is called the elastic range. Once the stress exceeds, however, a critical value, called the **yield strength**, s_y, some of the deformation of the material becomes permanent. In other words, the yield point separates the elastic region from the plastic range of materials. This is always important if one wants to know how large an applied stress needs to be in order for plastic deformation of a workpiece to occur. On the other hand, the yield strength provides the limit for how much a structural component can be stressed before unwanted permanent deformation takes place. As an example, a screwdriver has to have high yield strength; otherwise, it will deform upon application of a large twisting force.

The highest force (or stress) that a material can sustain is called the **tensile strength**, s_T (Figure 4.2), which is sometimes called *ultimate tensile strength* or *ultimate tensile stress*, s_{UTS}. At this point, a localized decrease in the cross-sectional area starts to occur. The material is said to undergo necking, as shown in Figure 4.3. Because the cross section is now reduced, a smaller force is needed to continue deformation until eventually the **breaking strength**, s_B, is reached (Figure 4.2).

The slope in the elastic part of

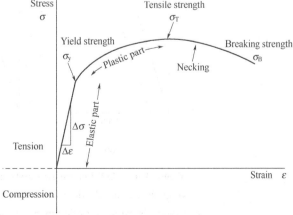

Figure 4.2 Schematic representation of a stress-strain diagram for a ductile materal

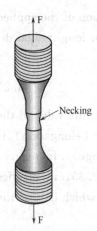

Figure 4.3 Necking of a test sample that was stressed in a tensile machine

the stress-strain diagram (Figure 4.2) is defined to be the **modulus of elasticity**, E, (or **Young's modulus**):

$$\frac{\sigma}{\varepsilon} = E \tag{4.3}$$

Equation 4.3 is generally referred to as **Hooke's Law**. For most typical metals the magnitude of this modulus ranges between 45 GPa for magnesium and 407 GPa for tungsten. For shear stress, t, Hooke's law is appropriately written as:

$$\frac{\tau}{\gamma} = G \tag{4.4}$$

where γ is the **shear strain** $\Delta a/a = \tan a \approx a$ and G is the **shear modulus.**

The modulus of elasticity is a parameter that reveals how "stiff" a material is, that is, it expresses the resistance of a material to elastic bending or elastic elongation. Specifically, a material having a large modulus and, therefore, a large slope in the stress-strain diagram deforms very little upon application of even a high stress. This material is said to have a high stiffness. This is always important if one requires close tolerances, such as for bearings, to prevent friction.

For metals, the point of yielding may not be determined precisely. As a consequence, a convention has been established wherein a straight line is constructed parallel to the elastic portion of the stress—strain curve at some specified strain offset, usually 0.002. This is demonstrated in Figure 4.4 (a).

Some steels and other materials exhibit the tensile stress—strain behavior as shown in

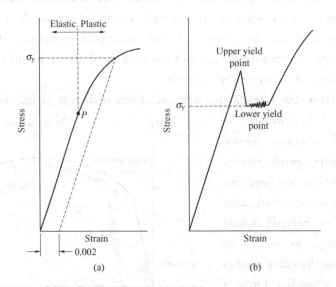

Figure 4.4 (a) Typical stress-strain behavior for a metal showing elastic and plastic deformations, the proportional limit P, and the yield strength as determined using the 0.002 strain offset method. (b) Representative stress-strain behavior found for some steels demonstrating the yield point phenomenon

Figure 4. 4 (b). The elastic-plastic transition is very well defined and occurs abruptly in what is termed a yield point phenomenon. At the upper yield point, plastic deformation is initiated with an actual decrease in stress. Continued deformation fluctuates slightly about some constant stress value, termed the lower yield point; stress subsequently rises with increasing strain. For metals that display this effect, the yield strength is taken as the average stress that is associated with the lower yield point, since it is well defined and relatively insensitive to the testing procedure. Thus, it is not necessary to employ the strain offset method for these materials.

Ductility is another important mechanical property. It is a measure of the degree of plastic deformation that has been sustained at fracture. A material that experiences very little or no plastic deformation upon fracture is termed *brittle*. The tensile stress-strain behaviors for both ductile and brittle materials are schematically illustrated in Figure 4. 5.

Ductility may be expressed quantitatively as either *percent elongation* or *percent reduction in area*. The percent elongation %EL is the percentage of plastic strain at fracture, or

$$\%EL = \left(\frac{l_f - l_0}{l_0}\right) \times 100 \quad (4.5)$$

where l_f is the fracture length and l_0 is the original gauge length as above. In as much as a significant proportion of the plastic deformation at fracture is confined to the neck region, the magnitude of %EL will depend on specimen gauge length. The shorter l_0, the greater is the fraction of total elongation from the neck and, consequently, the higher the value of %EL. Therefore, should be

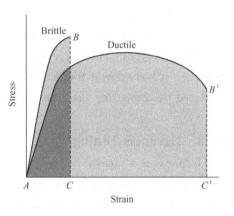

Figure 4. 5 Schematic representations of tensile stress-strain behavior for brittle and ductile materials loaded to fracture

specified when percent elongation values are cited; it is commonly 50 mm (2 in.). Percent reduction in area %RA is defined as

$$\%RA = \left(\frac{A_0 - A_f}{A_0}\right) \times 100 \quad (4.6)$$

where A_0 is the original cross-sectional area and A_f is the cross-sectional area at the point of fracture. Percent reduction in area values are independent of both l_0 and A_0. Furthermore, for a given material the magnitudes of %EL and %RA will, in general, be different. Most metals possess at least a moderate degree of ductility at room temperature; however, some become brittle as the temperature is lowered.

Knowledge of the ductility of materials is important for at least two reasons. First, it indicates to a designer the degree to which a structure will deform plastically before fracture. Second, it specifies the degree of allowable deformation during fabrication operations. We sometimes refer to relatively ductile materials as being "forgiving", in the sense that they may experience local deformation without fracture should there be an error in the magnitude of the design stress calculation. Brittle materials are *approximately* considered to be those having a fracture strain of less than about 5%. Thus, several important mechanical properties of metals may be determined from tensile stress-strain tests.

Toughness is a mechanical term that is used in several contexts; loosely speaking, it is a measure of the ability of a material to absorb energy up to fracture. Specimen geometry as well as the manner of load application is important in toughness determinations. For dynamic (high strain rate) loading conditions and when a notch (or point of stress concentration) is present, *notch toughness* is assessed by using an impact test. Furthermore, fracture toughness is a property indicative of a material's resistance to fracture when a crack is present.

For the static (low strain rate) situation, toughness may be ascertained from the results of a tensile stress-strain test. It is the area under the $\sigma\text{-}\epsilon$ curve up to the point of fracture. The units for toughness are the same as for resilience (i. e., energy per unit volume of material). For a material to be tough it must display both strength and ductility; often, ductile materials are tougher than brittle ones.

This is demonstrated in Figure 4.5, in which the stress-strain curves are plotted for both material types. Hence, even though the brittle material has higher yield and tensile strengths, it has a lower toughness than the ductile one, by virtue of lack of ductility; this is deduced by comparing the areas ABC and $A'B'C'$ in Figure 4.5.

4.1.3 Hardness Testing

Hardness is a measure of a material's resistance to localized plastic deformation. The tests employed assess differing combinations of the elastic, yielding and work-hardening characteristics. Quantitative hardness techniques have been developed over the years in which a small indenter is forced into the surface of a material to be tested, under controlled conditions of load and rate of application. The depth or size of the resulting indentation is measured, which in turn is related to a hardness number.

All the tests are essentially simple, inexpensive, and rapid to carry out and are virtually non-destructive, so they are well-suited as a means of quality control. The hardness of materials has been assessed by a wide variety of tests as followings.

The Brinell test

The surface of the material is indented by a hardened steel ball (whose diameter D is usually 10 mm) under a known load (L) (e. g. 3000 kg for steel) and the average diameter of the impression measured with a low-power microscope. The Brinell number (HB) which is in units of kgf/mm^2 is the ratio of the load to the contact surface area of th e indentation. Most machines have a set of charts for each loading force, from which the measured diameter is then converted to the appropriate HB number using a chart.

Semiautomatic techniques for measuring Brinell hardness are available. These employ optical scanning systems consisting of a digital camera mounted on a flexible probe, which allows positioning of the camera over the indentation. Data from the camera are transferred to a computer that analyzes the indentation, determines its size, and then calculates the Brinell hardness number.

The Vickers test

A diamond square-based pyramid of 136° angle is used as the indenter, which gives geometrically similar impressions under differing loads (which may range from 5 to 120 kg). A square indent is thus produced, and the user measures the average diagonal length and again reads the hardness number (HV) from the tables. The Brinell and Vickers hardness values are identical up to a hardness of about 300 kgf/mm^2, but distortion of the steel ball occurs in Brinell tests on hard materials, so that the test is not reliable above values of 600 kgf/mm^2.

For steels there is a useful empirical relationship between the UTS (in MPa) and HV (in kgf/mm^2), namely: UTS $\approx$ 3.2 HV.

The Rockwell test

Either a steel ball (Scale B) or a diamond cone (Scale C) is used and the indenter is first loaded with a minor load of 10 kgf, while the indicator for measuring the depth of the impression is set to zero. The appropriate major load is then applied and, after its removal, the dial gauge records the depth of the impression in terms of Rockwell numbers.

It should be pointed out that, in the case of materials which exhibit time-dependence of elastic modulus or yield stress (for example, most polymers), the size of the indentation will increase with time, so the hardness value will depend on the duration of application of the load.

The Knoop test

The Knoop test uses a diamond pyramidal indenter of apex angles 130° and 172.5°, thus giving a rhombohedral impression with one diagonal (L) being 7 × longer than the other and with a depth which is one thirtieth of L. It is particularly useful for measuring the relative hardness of brittle materials, such as glasses and ceramics, when lower loads (P) are employed than in the Vickers test. The Knoop Hardness Number (KHN, in kgf/mm^2) is given by the relation: $KHN = 14.229 P/L^2$.

Hardness Conversion

The facility to convert the hardness measured on one scale to that of another is most desirable. Hardness conversion data have been determined experimentally and found to be dependent on material type and characteristics. The most reliable conversion data exist for steels. Detailed conversion tables for various other metals and alloys are contained in ASTM Standard E 140, "Standard Hardness Conversion Tables for Metals".

4.2 Dislocations and Strengthening

4.2.1 The Role of Dislocations

We learned that the strength and the ductility of materials depend on the forces which hold the individual atoms together. This is, however, only one part of the story. Indeed, one can estimate the strength of a material taking solely the binding forces between the atoms into consideration, and compare this finding with experimental results. It is then observed that for ductile, pure materials the theoretical force which appears to be necessary to rupture a piece of metal by breaking the binding forces between atoms [Figure 4.6 (a)] would be between three and five orders of magnitude larger than those actually found in experiments. Moreover, the discrepancy between calculation and experiment varies for different crystal structures. Specifically, BCC materials are, as a rule, much stronger than FCC materials, whereas considerations based purely on atomic binding forces would predict approximately the same strengths.

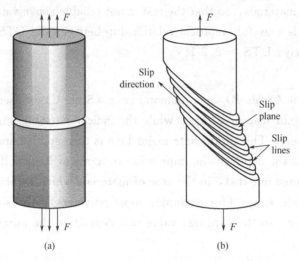

Figure 4.6 Schematic representation of two mechanisms by which a single crystal is assumed to be stretched when applying a force, F: (a) breaking inter-atomic bonds, and (b) considering slip

There exists, of course, an interrelationship between these binding forces and the plastic deformation of materials, but this interrelationship is rather complex. To better understand this, one has to know that crystalline materials deform predominantly by a process called **slip**, which occurs along specific lattice planes, called **slip planes**, as depicted for a single crystal rod in Figures 4.6 (b). The experimentally observed shear stress which needs to be applied in order that slip commences (called the critical resolved shear stress, t_0) is more than three orders of magnitude smaller than that which one would expect from calculations.

The source for this surprising discrepancy was ascribed in 1934 (by Orowan, Polanyi, and Taylor) to certain imperfections in crystals, called dislocations and their preferred movement (slip) along the slip planes. This concept initially met with considerable skepticism. It was not until the late 1950s that this notion was finally accepted, in particular once dislocations were seen in the transmission electron microscope.

Plastic deformation corresponds to the motion of large numbers of dislocations. From the microscopic aspect, the slip related with plastic deformation is produced by dislocation motion. The crystallographic plane along which the dislocation line traverses is the slip plane, as indicated in Figure 4.7. The local displacement (b) associated with a given dislocation is known as its **Burgers vector**.

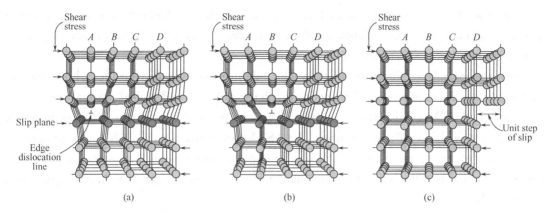

Figure 4.7 Atomic rearrangements that accompany the motion of an edge dislocation as it moves in response to an applied shear stress

All metals and alloys contain some dislocations that were introduced during solidification, during plastic deformation, and as a consequence of thermal stresses that result from rapid cooling. The number of dislocations, or **dislocation density** in a material, is expressed as the total dislocation length per unit volume or, equivalently, the number of dislocations that intersect a unit area of a random section. The units of dislocation density are millimeters of dislocation per cubic millimeter or just per square millimeter. Dislocation densities as low as 10^3 mm^{-2} are typically found in carefully solidified metal crystals. For heavily deformed metals, the density may run as high as 10^9 to 10^{10} mm^{-2}. Heat treating a deformed metal specimen can diminish the density to on the order of 10^5 to 10^6 mm^{-2}. By way of contrast, a typical dislocation density for ceramic materials is between 10^2 and 10^4 mm^{-2}.

4.2.2 Work Hardening

As a crystal is deformed, dislocations multiply and the dislocation density rises. The dislocations interact elastically with each other and the average spacing of the dislocation network decreases. The shear yield strength (t) of a crystal containing a network of dislocations of density ρ is given by:

$$\tau = \alpha G b (\rho^{1/2}), \tag{4.7}$$

where G is the shear modulus, b the dislocation Burgers vector, and α is a constant of value about 0.2.

As plastic deformation continues, therefore, the increase in dislocation density causes an increase in t— the well-known effect of work hardening.

As a technical means of producing a strong material, work hardening can only be employed in situations where large deformations are involved, such as in wire drawing and in the cold-rolling of sheet. This form of hardening is lost if the material is heated, because the additional thermal energy allows the dislocations to rearrange themselves, relaxing their stress fields through processes of recovery and being annihilated by recrystallization.

4.2.3 Grain Size Strengthening

Metals are usually used in polycrystalline form and, thus, dislocations are unable to move long distances without being held up at grain boundaries. Metal grains are not uniform in shape and size—their three-dimensional structure resembles that of a soap froth. There are two main methods of measuring the grain size:

(a) the mean linear intercept method which defines the average chord length intersected by the grains on a random straight line in the planar polished and etched section, and

(b) the ASTM comparative method, in which standard charts of an idealized network are compared with the microstructure. The ASTM grain size number (N) is related to n, the number of grains per square inch in the microsection observed at a magnification of $100\times$. The smaller the average grain diameter (d), the higher the ASTM grain size number, N.

The tensile yield strength (s_y) of polycrystals is higher the smaller the grain size, these parameters being related through the *Hall-Petch* equation:

$$\sigma_y = \sigma_0 + k_y d^{-1/2}, \tag{4.8}$$

Where k_y is a material constant and s_0 is the yield stress of a single crystal of similar composition and dislocation density.

4.2.4 Alloy Hardening

Work hardening and grain-size strengthening, which we have considered so far, can be applied to a pure metal. The possibility of changing the composition of the material by alloying presents further means of strengthening. We will consider two ways in which alloying elements may be used to produce strong materials: solute hardening and precipitation hardening.

Solute hardening

We have learned that two types of solid solution may be formed, namely interstitial and substitutional solutions. The presence of a "foreign" atom in the lattice will give rise to local stresses which will impede the movement of dislocations, hence raising the yield stress of the solid.

This effect is known as **solute hardening**, and its magnitude will depend on the concentration of solute atoms in the alloy and also upon the magnitude of the local misfit strains associated with the individual solute atoms. It is also recognized that the solubility of an element in a given crystal is itself dependent upon the degree of misfit—indeed, if the atomic sizes of the solute and solvent differ by more than about 14%, then only very limited solid solubility occurs. There must thus be a compromise between these two effects in a successful solution-hardened material—i.e., there must be sufficient atomic misfit to give rise to local lattice strains, but there must also be appreciable solubility.

Precipitation hardening

Thermal treatment can be used to control the size and distribution of second-phase particles in any alloy which undergoes a phase transformation in the solid state. In many alloy systems, the solid solubility changes with temperature. Above a certain temperature, a single phase (α) solid solution exists and, if the material is quenched rapidly from this temperature range, a supersaturated α solid solution is formed. If the temperature is then raised again in order to allow solid state diffusion to proceed, the supersaturation will be relieved by the nucleation and growth of a precipitated second phase.

In alloys of relatively low melting-point (in aluminum alloys, for example), there will be an appreciable diffusion rate of solute atoms at room temperature, so that over a period of time a second phase will precipitate out in a very finely divided form. This effect is known as "**aging**", but, in most alloys, the temperature has to be raised in order to cause precipitation to occur and the material is said to be "artificially aged". Low ageing temperatures correspond to high supersaturation and prolific nucleation of precipitates occurs, whereas at higher ageing temperatures (lower supersaturation) fewer, coarser particles are formed.

Quenching and ageing therefore constitute a very powerful means of controlling the distribution of a precipitate of second phase in an alloy. These precipitates can have a profound effect upon the mobility of dislocations and it is possible to produce large changes in the yield strength of such alloys by suitable heat-treatment. A great advantage is that the required strength can be induced in a product at the most convenient stage in its manufacture. For example, the alloy may be retained in a soft form throughout the period when it is being shaped by forging and it is finally hardened by precipitation in order to give it good strength in service.

On being held up by a precipitate, a dislocation can continue in its path across the crystal in two possible ways. If the particles are very close together, the dislocation may **cut through** each particle, but if the particles are further apart, the dislocation may **loop between** the particles. During the ageing process, as the particles grow, the stress increment required to make the dislocations cut them also rises. As ageing proceeds, the particles gradually increase in size and, because they are fewer in number, the average spacing between the particles also increases. The stress increment to cause dislocation looping decreases as the interparticle spacing increases.

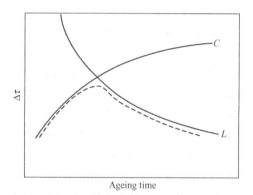

Figure 4.8 Showing change in yield stress ($\Delta\tau$) with aging time for a precipitation-hardened alloy. Curve C is followed if the precipitates are cut by dislocations, and curve L is followed if the dislocations loop between the particles. The response is given by the dashed curve

As ageing continues, the measured yield stress would therefore be expected to follow the form of the dotted curve in Figure 4.8 and this general pattern of behavior, with an optimum ageing time to give a maximum hardness, is commonly observed in many commercial alloys. The time to peak hardness depends on the solute diffusion rate and, thus, on the ageing temperature.

4.3 Failure

4.3.1 Introduction

Despite the great strides forward that have been made in technology, failures continue to occur, often accompanied by great human and economic loss. Since prehistoric times, failures have often resulted in taking one step back and two steps forward, but with severe consequences for the designers and builders. For example, according to the Code of Hammurabi, "If a builder build a house for a man and do not make its construction firm, and the house which he has built collapse and cause the death of the owner of the house, that builder shall be put to death".

The failure of bridges, viaducts, cathedrals, and so on, resulted in better designs, better materials, and better construction procedures. Mechanical devices, such as wheels and axles, were improved through empirical insights gained through experience, and these improvements often worked out quite well. An example of an evolved design that did not work out well is related to the earthquake that struck Kobe, Japan, in 1995. That area of Japan had been free of damaging earthquakes for some time, but had been visited frequently by typhoons. To stabilize homes against the ravages of typhoons, the local building practice was to use a rather heavy roof structure. Unfortunately, when the earthquake struck, the collapse of these heavy roofs caused considerable loss of life as well as property damage. The current design codes for this area have been revised to reflect a concern for both typhoons and earthquakes.

The designs of commonplace products have often evolved rapidly to make them safer. For example, consider the carbonated soft-drink bottle cap. At one time, a metal cap was firmly crimped to a glass bottle, requiring a bottle opener for removal (still used for beer). Then came the easy-opening, twist-off metal cap. These caps were made of a thin, circular piece of aluminum that was shaped by a tool at the bottling plant to conform to the threads of the glass bottle (still used for some liquor). If the threads were worn, or if the shaping tool did not maintain proper alignment, then the connection between cap and bottle would be weak and the cap might spontaneously blow off the bottle on the supermarket shelf. Worse than that, there were a number of cases where, during the twisting-off process, the expanding gas suddenly propelled a weakly attached cap from the bottle and caused eye damage. To guard against this danger, the metal caps were redesigned to have a series of closely spaced perforations along the upper side of the cap, so that as the seal between the cap and bottle was broken at the start of the twisting action, the gas pressure was vented, and the possibility of causing an eye injury was minimized. The next stage in the evolution of bottle cap design has been to use plastic bottles and plastic caps. In a current design, the threads on the plastic bottle are slotted, so that, as in the case of the perforated metal cap, as the cap is twisted the CO_2 gas is vented, and the danger of causing eye damage is reduced.

The design of a component or structure often calls upon the engineer to minimize the possibility of failure. Thus, it is important to understand the mechanics of the various failure modes—i.e., fracture, fatigue, and creep—and, it is the responsibility of the engineer to anticipate and plan for possible failure and, in the event that failure does occur, to assess its cause and then take appropriate preventive measures against future incidents.

4.3.2 Fundamentals of Fracture

Simple fracture is the separation of a body into two or more pieces in response to an imposed static stress and at temperatures that are low relative to the melting temperature of the material. For engineering materials, two fracture modes are possible: **ductile** and **brittle.** Classification is based on the ability of a material to experience plastic deformation. Ductile materials typically exhibit substantial plastic deformation with high energy absorption before fracture. On the other hand, there is normally little or no plastic deformation with low energy absorption accompanying a brittle fracture.

Any fracture process involves two steps—crack formation and propagation—in response to an imposed stress. The mode of fracture is highly dependent on the mechanism of crack propagation. Ductile fracture is characterized by extensive plastic deformation in the vicinity of an advancing crack. Furthermore, the process proceeds relatively slowly as the crack length is extended. Such a crack is often said to be stable. That is, it resists any further extension unless there is an increase in the applied stress. In addition, there will ordinarily be evidence of appreciable gross deformation at the fracture surfaces. On the other hand, for brittle fracture, cracks may spread extremely rapidly, with very little accompanying plastic deformation. Such cracks may be said to be unstable, and crack propagation, once started, will continue spontaneously without an increase in magnitude of the applied stress.

Ductile fracture is almost always preferred for two reasons. First, brittle fracture occurs suddenly and catastrophically without any warning; this is a consequence of the spontaneous and rapid crack propagation. On the other hand, for ductile fracture, the presence of plastic deformation gives warning that fracture is imminent, allowing preventive measures to be taken. Second, more strain energy is required to induce ductile fracture inasmuch as ductile materials are generally tougher. Under the action of an applied tensile stress, most metal alloys are ductile, whereas ceramics are notably brittle, and polymers may exhibit both types of fracture.

4.3.3 Ductile Fracture

Ductile fracture surfaces will have their own distinctive features on both macroscopic and microscopic levels. Figure 4.9 shows schematic representations for two characteristic macroscopic fracture profiles. The configuration shown in Figure 4.9 (a) is found for extremely soft metals, such as pure gold and lead at room temperature, and other metals, polymers, and inorganic glasses at elevated temperatures. These highly ductile materials neck down to a point fracture,

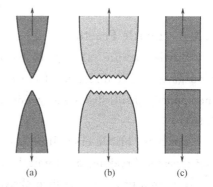

Figure 4.9 (a) Highly ductile fracture in which the specimen necks down to a point. (b) Moderately ductile fracture after some necking. (c) Brittle fracture without any plastic deformation

showing virtually 100% reduction in area.

The most common type of tensile fracture profile for ductile metals is that represented in Figure 4. 9 (b), where fracture is preceded by only a moderate amount of necking.

Much more detailed information regarding the mechanism of fracture is available from microscopic examination. When the fibrous central region of a fracture surface like Figure 4. 9 (b) is examined with the electron microscope at a high magnification, it will be found to consist of numerous spherical "dimples" (Figure 4. 10); this structure is characteristic of fracture resulting from uniaxial tensile failure.

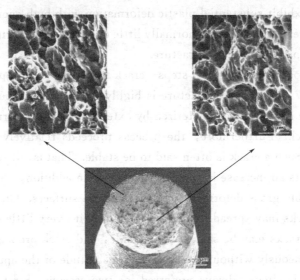

Figure 4. 10 Scanning electron micrographs at low magnification (center) and high magnification (right and left) of AISI 1008 steel specimen ruptured in tension. Notice the equiaxal dimples in the central region and elongated dimples on the shear walls, the sides of the cup

4. 3. 4 Brittle Fracture

Brittle fracture takes place without any appreciable deformation, and by rapid crack propagation. The direction of crack motion is very nearly perpendicular to the direction of the applied tensile stress and yields a relatively flat fracture surface, as indicated in Figure 4. 9 (c).

For most brittle crystalline materials, crack propagation corresponds to the successive and repeated breaking of atomic bonds along specific crystallographic planes [Figure 4. 11 (a)]; such a process is termed **cleavage**. This type of fracture is said to be **transgranular** (or transcrystalline), because the fracture cracks pass through the grains. This cleavage feature is shown at a higher magnification in the scanning electron micrograph of Figure 4. 11 (b).

In soe alloys, crack propagation is along grain boundaries [Figure 4. 12 (a)]; this fracture is termed **intergranular**. Figure 4. 12 (b) is a scanning electron micrograph showing a typical intergranular fracture, in which the three-dimensional nature of the grains may be seen. This type of fracture normally results subsequent to the occurrence of processes that weaken or embrittle grain boundary regions.

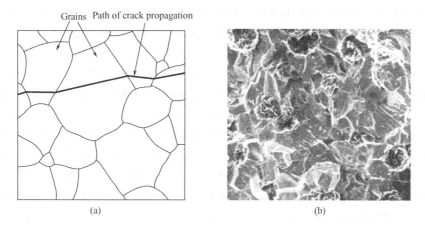

Figure 4.11 (a) Schematic cross-section profile showing crack propagation through the interior of grains for transgranular fracture. (b) Scanning electron fractograph of ductile cast iron showing a transgranular fracture surface. Magnification unknown

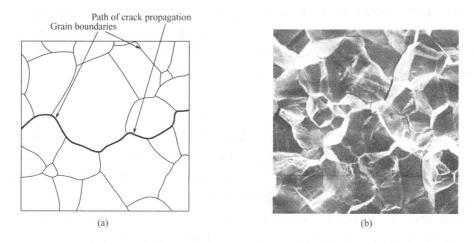

Figure 4.12 (a) Schematic cross-section profile showing crack propagation along grain boundaries for intergranular fracture. (b) Scanning electron fractograph showing an intergranular fracture surface. 50×

4.3.5 Fracture Mechanics in Design

There are two related material parameters that define a material's ability to resist the crack propagation that results in material fracture. One is the fracture toughness, and the other is the toughness. The use of these two parameters in design is an outgrowth of the development of the engineering science of fracture mechanics. This development was spurred by incidents of Liberty Ship that broke in half while docked. The Liberty Ships were constructed in response to the need for ocean transport vessels in World War II. Construction of conventional transport ships took considerable time. A primary reason was that the hull of the conventional transports consisted of a large number of individual sections riveted to the main ship support structure; placement of the individual sections was time consuming. The design for the Liberty Ships invoked an all-welded hull. That is the individual hull sections were welded together rather than being mechanically joined. The time to construct a transport ship with a welded hull was much less.

There was one—one very major—problem with the welded-hull construction. Specifically, an advancing crack in such a hull could propagate through the whole of it; thus, the catastrophe that the ship fractured at the middle at pier occurred. Conventional transport ships did not suffer similarly. A crack running in them would generally be "arrested" at the juncture between individual hull sections. The Liberty Ship problem spurred intense research, first at the United States Naval Research Laboratory, where George Irwin and a number of his colleagues played a pivotal role in developing the engineering science of fracture mechanics, which permits design against fracture, just as a tensile test permits design against permanent deformation.

The design philosophy of fracture mechanics is that the stress to which a material is subjected must be less than the stress required to fracture it. That sounds simple enough; application is more complicated. In particular, the fracture stress is that stress needed to propagate cracks, such cracks being assumed initially present in the material or having developed in it during service. Indeed, under the worst circumstances, engineered structures may contain cracks of macroscopic dimensions. Depending on flaw severity and material properties, a material that undergoes ductile tensile fracture may fail in a macroscopically brittle fashion.

4.3.6　Fracture Toughness

The toughness of a material must be distinguished from its ductility. It is true that ductile materials are frequently tough, but toughness combines both strength and ductility, so that some soft metals like lead are too weak to be tough whereas glass-reinforced plastics are very tough although they exhibit little plastic strain.

One approach to toughness measurement is to measure the work done in breaking a specimen of the material, such as in the Charpy-type of impact test. Here a bar of material is broken by a swinging pendulum and the energy lost by the pendulum in breaking the sample is obtained from the height of the swing after the sample is broken. A serious disadvantage of such tests is the difficulty of reproducibility of the experimental conditions by different investigators, so that impact tests can rarely be scaled up from laboratory to service conditions and the data obtained cannot be considered to be true material parameters.

Fracture toughness is now assessed by establishing the conditions under which a sharp crack will begin to propagate through the material and a number of interrelated parameters may be employed to express this property. An expression has been developed that relates this critical stress for crack propagation (s_C) and crack length (a) as

$$K_C = Y s_C \sqrt{pa} \tag{4.9}$$

In this expression K_C is the fracture toughness, a property that is a measure of a material's resistance to brittle fracture when a crack is present. Worth noting is that has the unusual units ofMPa $\sqrt{m}$. Furthermore, Y is a dimensionless parameter or function that depends on both crack and specimen sizes and geometries, as well as the manner of load application. For planar specimens containing cracks that are much shorter than the specimen width, Y has a value of approximately unity.

4.3.7 Fatigue

Fatigue is a form of failure that occurs in structures subjected to dynamic and fluctuating stresses (e.g., bridges, aircraft, and machine components). Under these circumstances it is possible for failure to occur at a stress level considerably lower than the tensile or yield strength for a static load. The term "fatigue" is used because this type of failure normally occurs after a lengthy period of repeated stress or strain cycling. Fatigue is important inasmuch as it is the single largest cause of failure in metals, estimated to comprise approximately 90% of all metallic failures; polymers and ceramics (except for glasses) are also susceptible to this type of failure. Furthermore, fatigue is catastrophic and insidious, occurring very suddenly and without warning.

Fatigue failure is brittle-like in nature even in normally ductile metals, in that there is very little, if any, gross plastic deformation associated with failure. The process occurs by the initiation and propagation of cracks, and ordinarily the fracture surface is perpendicular to the direction of an applied tensile stress.

The process of fatigue failure is characterized by three distinct steps: ①crack initiation, wherein a small crack forms at some point of high stress concentration; ②crack propagation, during which this crack advances incrementally with each stress cycle; and ③final failure, which occurs very rapidly once the advancing crack has reached a critical size. Cracks associated with fatigue failure almost always initiate (or nucleate) on the surface of a component at some point of stress concentration. Crack nucleation sites include surface scratches, sharp fillets, keyways, threads, dents, and the like. In addition, cyclic loading can produce microscopic surface discontinuities resulting from dislocation slip steps that may also act as stress raisers, and therefore as crack initiation sites.

4.3.8 Creep

Materials are often placed in service at elevated temperatures and exposed to static mechanical stresses (e.g., turbine rotors in jet engines and steam generators that experience centrifugal stresses, and high-pressure steam lines). Deformation under such circumstances is termed **creep**. Defined as the time-dependent and permanent deformation of materials when subjected to a constant load or stress, creep is normally an undesirable phenomenon and is often the limiting factor in the lifetime of a part. It is observed in all materials types; for metals it becomes important only for temperatures greater than about 0.4 T_m (T_m = absolute melting temperature). Amorphous polymers, which include plastics and rubbers, are especially sensitive to creep deformation.

There are several factors that affect the creep characteristics of metals. These include melting temperature, elastic modulus, and grain size. In general, the higher the melting temperature, the greater the elastic modulus, and the larger the grain size, the better is a material's resistance to creep. Relative to grain size, smaller grains permit more grain-boundary sliding, which results in higher creep rates. This effect may be contrasted to the influence of grain size on the mechanical behavior at low temperatures (i.e., increase in both strength and toughness).

Stainless steels, the refractory metals, and the superalloys are especially resilient to creep and are commonly employed in high temperature service applications. The creep resistance of the cobalt and nickel superalloys is enhanced by solid-solution alloying, and also by the addition of a dispersed phase that is virtually insoluble in the matrix. In addition, advanced processing techniques have been utilized; one such technique is directional solidification, which produces either highly elongated grains or single-crystal components. Another is the controlled unidirectional solidification of alloys having specially designed compositions wherein two-phase composites result.

4.4 Phase Diagrams and Phase Transformations in Metals

4.4.1 Introduction

The understanding of phase diagrams for alloy systems is extremely important because there is a strong correlation between microstructure and mechanical properties, and the development of microstructure of an alloy is related to the characteristics of its phase diagram. In addition, phase diagrams provide valuable information about melting, casting, crystallization, and other phenomena.

It is necessary to establish a foundation of definitions and basic concepts relating to alloys, phases, and equilibrium before delving into the interpretation and utilization of phase diagrams. The term **component** is frequently used in this discussion; components are pure metals and/or compounds of which an alloy is composed. For example, in a copper-zinc brass, the components are Cu and Zn. Solute and solvent, which are also common terms. Another term used in this context is **system**, which has two meanings. First, "system" may refer to a specific body of material under consideration (e.g., a ladle of molten steel). Or it may relate to the series of possible alloys consisting of the same components, but without regard to alloy composition (e.g., the iron-carbon system). For many alloy systems and at some specific temperature, there is a maximum concentration of solute atoms that may dissolve in the solvent to form a solid solution; this is called a solubility limit. The addition of solute in excess of this solubility limit results in the formation of another solid solution or compound that has a distinctly different composition. To illustrate this concept, consider the sugar-water system. Initially, as sugar is added to water, a sugar-water solution or syrup forms. As more sugar is introduced, the solution becomes more concentrated, until the solubility limit is reached, or the solution becomes saturated with sugar. At this time the solution is not capable of dissolving any more sugar, and further additions simply settle to the bottom of the container. Thus, the system now consists of two separate substances: a sugar-water syrup liquid solution and solid crystals of undissolved sugar.

Also critical to the understanding of phase diagrams is the concept of a **phase**. A phase may be defined as a homogeneous portion of a system that has uniform physical and chemical characteristics. Every pure material is considered to be a phase; so also is every solid, liquid, and gaseous solution. If more than one phase is present in a given system, each will have its own distinct properties, and a boundary separating the phases will exist across which there will be a discontinuous and abrupt change in physical and/or chemical characteristics. When two phases are present in a system, it is not necessary that there be a difference in both physical and chemical properties; a disparity in one or the other set of properties is sufficient. When water and ice are present in a container, two separate phases exist; they are physically dissimilar (one is a solid, the other is a liquid) but identical in chemical makeup. Also, when a substance can exist in two or more polymorphic forms (e. g., having both FCC and BCC structures), each of these structures is a separate phase because their respective physical characteristics differ.

Sometimes, a single-phase system is termed "homogeneous". Systems composed of two or more phases are termed "mixtures" or "heterogeneous systems". Most metallic alloys and, for that matter, ceramic, polymeric, and composite systems are heterogeneous. Ordinarily, the phases interact in such a way that the property combination of the multiphase system is different from, and more attractive than, either of the individual phases.

Equilibrium is another essential concept that is best described in terms of a thermodynamic quantity called the free energy. In brief, free energy is a function of the internal energy of a system, and also the randomness or disorder of the atoms or molecules (or entropy). A system is at equilibrium if its free energy is at a minimum under some specified combination of temperature, pressure, and composition. In a macroscopic sense, this means that the characteristics of the system do not change with time but persist indefinitely; that is, the system is stable. A change in temperature, pressure, and/or composition for a system in equilibrium will result in an increase in the free energy and in a possible spontaneous change to another state whereby the free energy is lowered.

4.4.2 Phase Diagrams

Another type of extremely common phase diagram is one in which temperature and composition are variable parameters, and pressure is held constant-normally 1 atm. There are several different varieties; in the present discussion, we will concern ourselves with binary alloys-those that contain two components. If more than two components are present, phase diagrams become extremely complicated and difficult to represent. An explanation of the principles governing and the interpretation of phase diagrams can be demonstrated using binary alloys even though most alloys contain more than two components.

Binary isomorphous systems

Possibly the easiest type of binary phase diagram to understand and interpret is the type that is characterized by the copper-nickel system [Figure 4.13 (a)]. Temperature is plotted along the ordinate, and the abscissa represents the composition of the alloy, in weight percent (bottom) and atom percent (top) of nickel. The composition ranges from 0 wt% Ni (100 wt% Cu) on the left horizontal extremity to 100 wt% Ni (0 wt% Cu) on the right. Three different phase regions, or fields, appear on the diagram, an alpha (α) field, a liquid (L) field, and a two-phase field. Each region is defined by the phase or phases that exist over the range of temperatures and compositions delimited by the phase boundary lines.

The liquid L is a homogeneous liquid solution composed of both copper and nickel. The α phase is a substitutional solid solution consisting of both Cu and Ni atoms, and having an FCC crystal structure. At temperatures below about copper and nickel are mutually soluble in each other in the solid state for all compositions. The copper-nickel system is termed **isomorphous** because of this complete liquid and solid solubility of the two components.

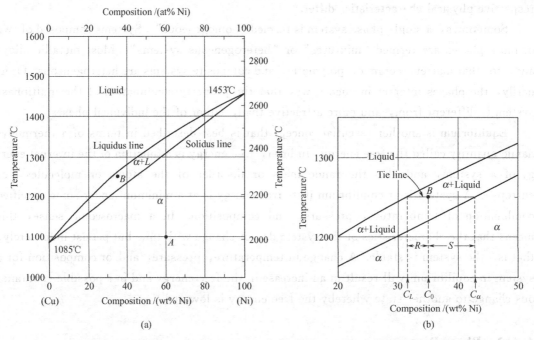

Figure 4.13 (a) The copper-nickel phase diagram, (b) A portion of the copper-nickel phase diagram for which compositions and phase amounts are determined at point B

The relative amounts (as fraction or as percentage) of the phases present at equilibrium can be computed with the aid of phase diagrams. If the composition and temperature position is located within a two-phase region, the tie line must be utilized in conjunction with a procedure that is often called the lever rule.

Consider the example shown in Figure 4.13 (b), in which at 1250°C both α and liquid phases are present for a 35 wt% Ni-65 wt% Cu alloy. The problem is to compute the fraction of each of the α and liquid phases. The tie line has been constructed that was used for the determination of α and L phase compositions. Let the overall alloy composition be located along the tie line and denoted as C_0 and mass fractions be represented by W_L and W_α for the respective phases. From the lever rule, W_L may be computed according to

$$W_L = \frac{S}{R+S} = \frac{C_a - C_0}{C_a - C_L}. \tag{4.10}$$

Similarly, for the α phase,

$$W_a = \frac{R}{R+S} = \frac{C_L - C_0}{C_a - C_L}. \tag{4.11}$$

Thus, the lever rule may be employed to determine the relative amounts or fractions of phases in any two-phase region for a binary alloy if the temperature and composition are known and if equilibrium has been established. Its derivation is presented as an example problem.

Binary eutectic systems

Another type of common and relatively simple phase diagram found for binary alloys is shown in Figure 4.14 for the copper-silver system; this is known as a binary eutectic phase diagram. A number of features of this phase diagram are important and worth noting. First, three single-phase regions are found on the diagram: α, β, and liquid. The a phase is a solid solution rich in copper; it has silver as the solute component and an FCC crystal structure. The β-phase solid solution also has an FCC structure, but copper is the solute. Pure copper and pure silver are also considered to be α and β phases, respectively.

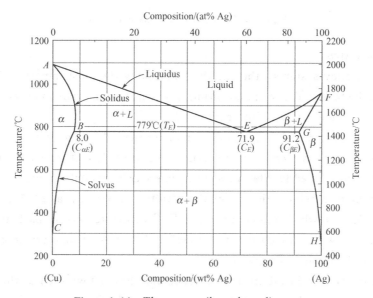

Figure 4.14 The copper-silver phase diagram

An important reaction occurs for an alloy of composition C_E as it changes temperature in passing through T_E; this reaction may be written as follows:

$$L(C_E) \underset{\text{heating}}{\overset{\text{cooling}}{\rightleftharpoons}} \alpha(C_{\alpha E}) + \beta(C_{\beta E}) \tag{4.12}$$

This is called a **eutectic reaction** (eutectic means easily melted), and C_E and T_E represent the eutectic composition and temperature, respectively.

Another common eutectic system is that for lead and tin; the phase diagram (Figure 4.15) has a general shape similar to that for copper-silver. For the lead-tin system the solid solution phases are also designated by α and β; in this case, α represents a solid solution of tin in lead and, for β, tin is the solvent and lead is the solute. The eutectic invariant point is located at 61.9 wt% Sn and 183 ℃.

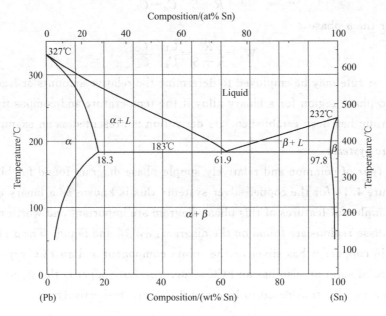

Figure 4.15 The lead-tin phase diagram

The iron-carbon system

Of all binary alloy systems, the one that is possibly the most important is that for iron and carbon. Both steels and cast irons, primary structural materials in every technologically advanced culture, are essentially iron-carbon alloys.

A portion of the iron-carbon phase diagram is presented in Figure 4.15. Pure iron, upon heating, experiences two changes in crystal structure before it melts. At room temperature the stable form, called **ferrite**, or α iron, has a BCC crystal structure. Ferrite experiences a polymorphic transformation to FCC **austenite**, or γ iron, at 912℃. This austenite persists to 1394℃, at which temperature the FCC austenite reverts back to a BCC phase known as d ferrite, which finally melts at 1538℃. All these change are apparent along the left vertical axis of the phase diagram.

The composition axis in Figure 4.16 extends only to 6.70 wt% C; at this concentration the intermediate compound iron carbide, or **cementite** (Fe_3C), is formed, which is represented by a vertical line on the phase diagram. Thus, the iron-carbon system may be divided into two parts: an iron-rich portion, as in Figure 4.15, and the other (not shown) for compositions between 6.70 and 100 wt% C (pure graphite). In practice, all steels and cast irons have carbon contents less than 6.70 wt% C; therefore, we consider only the iron-iron carbide system. Figure 4.15 would be more appropriately labeled the $Fe-Fe_3C$ phase diagram, since Fe_3C is now considered to be a component.

In the BCC α ferrite, only small concentrations of carbon are soluble; the maximum solubility is 0.022 wt% at 727℃. This particular iron-carbon phase is relatively soft, may be made magnetic at temperatures below 768℃. The austenite, or γ phase of iron, when alloyed with carbon alone, is not stable below 727℃. The maximum solubility of carbon in austenite, 2.14 wt%, occurs at 1147℃. The d ferrite is virtually the same as α ferrite, except for the range of temperatures over which each exists. Cementite (Fe_3C) forms when the solubility limit of carbon in α ferrite is exceeded below 727℃. Mechanically, cementite is very hard and brittle; the strength of some steels is greatly enhanced by its presence.

Strictly speaking, cementite is only metastable; that is, it will remain as a compound indefinitely at room temperature. Thus, the phase diagram in Figure 4.16 is not a true equilibrium one because cementite is not an equilibrium compound. However, inasmuch as the decomposition rate of cementite is extremely sluggish, virtually all the carbon in steel will be as Fe_3C instead of graphite, and the iron-iron carbide phase diagram is, for all practical purposes, valid.

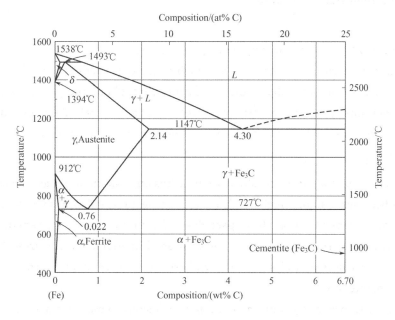

Figure 4.16 The iron-iron carbide phase diagram

4.4.3 Phase Transformations

A variety of phase transformations are important in the processing of materials, and usually they involve some alteration of the microstructure. These transformations are divided into three classifications. In one group are simple diffusion-dependent transformations in which there is no change in either the number or composition of the phases present. These include solidification of a pure metal, allotropic transformations, and, recrystallization and grain growth. In another type of diffusion-dependent transformation, there is some alteration in phase compositions and often in the number of phases present; the final microstructure ordinarily consists of two phases. The eutectoid reaction, described by Equation 4.10, is of this type. The third kind of transformation is diffusionless, wherein a metastable phase is produced. A martensitic transformation, which may be induced in some steel alloys, falls into this category.

With phase transformations, normally at least one new phase is formed that has different physical/chemical characteristics and/or a different structure than the parent phase. Furthermore, most phase transformations do not occur instantaneously. Rather, they begin by the formation of numerous small particles of the new phase (s), which increase in size until the transformation has reached completion. The progress of a phase transformation may be broken down into two distinct stages: **nucleation** and **growth**. Nucleation involves the appearance of very small particles, or nuclei of the new phase (often consisting of only a few hundred atoms), which are capable of growing. During the growth stage these nuclei increase in size, which results in the disappearance of some (or all) of the parent phase. The transformation reaches completion if the growth of these new phase particles is allowed to proceed until the equilibrium fraction is attained.

Nucleation

There are two types of nucleation: homogeneous and heterogeneous. The distinction between them is made according to the site at which nucleating events occur. For the homogeneous type, nuclei of the new phase form uniformly throughout the parent phase, whereas for the heterogeneous type, nuclei form preferentially at structural inhomogeneities, such as container surfaces, insoluble impurities, grain boundaries, dislocations, and so on.

Growth

The growth step in a phase transformation begins once an embryo has exceeded the critical size, and becomes a stable nucleus. Note that nucleation will continue to occur simultaneously with growth of the new phase particles; of course, nucleation cannot occur in regions that have already transformed to the new phase. Furthermore, the growth process will cease in any region where particles of the new phase meet, since here the transformation will have reached completion.

Particle growth occurs by long-range atomic diffusion, which normally involves several steps—for example, diffusion through the parent phase, across a phase boundary, and then into the nucleus. Consequently, the growth rate is determined by the rate of diffusion, and its temperature dependence is the same as for the diffusion coefficient.

4.5 Applications and Processing of Metal Alloys

4.5.1 Introduction

Often a materials problem is really one of selecting the material that has the right combination of characteristics for a specific application. Therefore, the people who are involved in the decision making should have some knowledge of the available options. Materials selection decisions may also be influenced by the ease with which metal alloys may be formed or manufactured into useful components. Alloy properties are altered by fabrication processes, and, in addition, further property alterations may be induced by the employment of appropriate heat treatments.

4.5.2 Types of Metal Alloys

As mentioned at the beginning of this chapter, ferrous alloys, those in which iron is the principal constituent, include steels and cast irons. These alloys and their characteristics are the first topics of discussion of this section. The nonferrous ones—all alloys that are not iron based—are treated next.

Steels

Steels are iron-carbon alloys that may contain appreciable concentrations of other alloying elements; there are thousands of alloys that have different compositions and/or heat treatments. The mechanical properties are sensitive to the content of carbon, which is normally less than 1.0 wt%. Some of the more common steels are classified according to carbon concentration-namely, into low-, medium-, and high-carbon types. Subclasses also exist within each group according to the concentration of other alloying elements. **Plain carbon steels** contain only residual concentrations of impurities other than carbon and a little manganese. For **alloy steels**, more alloying elements are intentionally added in specific concentrations.

Low-carbon steels generally contain less than about 0.25 wt% C and are unresponsive to heat treatments intended to form martensite; strengthening is accomplished by cold work. Microstructures consist of ferrite and pearlite constituents. As a consequence, these alloys are relatively soft and weak but have outstanding ductility and toughness; in addition, they are machinable, weldable, and, of all steels, are the least expensive to produce. Typical applications include automobile body components, structural shapes (I-beams, channel and angle iron), and sheets that are used in pipelines, buildings, bridges, and tin cans. Another group of low-carbon alloys are the high-strength, low-alloy (HSLA) steels. They contain other alloying elements such as copper, vanadium, nickel, and molybdenum in combined concentrations as high as 10 wt%, and possess higher strengths than the plain low-carbon steels. In normal atmospheres, the HSLA steels are more resistant to corrosion than the plain carbon steels, which they have replaced in many applications where structural strength is critical (e.g., bridges, towers, support columns in high-rise buildings, and pressure vessels).

The medium-carbon steels have carbon concentrations between about 0.25 and 0.60 wt%. These alloys may be heat treated by austenitizing, quenching, and then tempering to improve their mechanical properties. These heat-treated alloys are stronger than the low-carbon steels, but at a sacrifice of ductility and toughness. Applications include railway wheels and tracks, gears, crankshafts, and other machine parts and high-strength structural components calling for a combination of high strength, wear resistance, and toughness.

The high-carbon steels, normally having carbon contents between 0.60 and 1.4 wt%, are the hardest, strongest, and yet least ductile of the carbon steels. They are almost always used in a hardened and tempered condition and, as such, are especially wear resistant and capable of holding a sharp cutting edge. The tool and die steels are high-carbon alloys, usually containing chromium, vanadium, tungsten, and molybdenum. These steels are utilized as cutting tools and dies for forming and shaping materials, as well as in knives, razors, hacksaw blades, springs, and high-strength wire.

The stainless steels are highly resistant to corrosion (rusting) in a variety of environments, especially the ambient atmosphere. Their predominant alloying element is chromium; a concentration of at least 11 wt% Cr is required. Corrosion resistance may also be enhanced by nickel and molybdenum additions. A wide range of mechanical properties combined with excellent resistance to corrosion make stainless steels very versatile in their applicability.

Cast irons

Generically, cast irons are a class of ferrous alloys with carbon contents above 2.14 wt%; in practice, however, most cast irons contain between 3.0 and 4.5 wt% C and, in addition, other alloying elements. Compared to steels, they are easily melted and amenable to casting. Furthermore, some cast irons are very brittle, and casting is the most convenient fabrication technique. For most cast irons, the carbon exists as graphite, and both microstructure and mechanical behavior depend on composition and heat treatment. The most common cast iron types are gray, nodular, white, malleable, and compacted graphite.

The carbon and silicon contents of **gray cast irons** vary between 2.5 and 4.0 wt% and 1.0 and 3.0 wt%, respectively. Mechanically, gray iron is comparatively weak and brittle in tension. Strength and ductility are much higher under compressive loads. Gray irons do have some desirable characteristics and, in fact, are utilized extensively. They are very effective in damping vibrational energy. Base structures for machines and heavy equipment that are exposed to vibrations are frequently constructed of this material.

Adding a small amount of magnesium and/or cerium to the gray iron before casting produces a distinctly different microstructure and set of mechanical properties. Graphite still forms, but as nodules or sphere-like particles instead of flakes. The resulting alloy is called **nodular iron**. Castings are stronger and much more ductile than gray iron. Typical applications for this material include valves, pump bodies, crankshafts, gears, and other automotive and machine components.

For low-silicon cast irons (containing less than 1.0 wt% Si) and rapid cooling rates, most of the carbon exists as cementite instead of graphite. A fracture surface of this alloy has a white appearance, and thus it is termed **white cast iron**. As a consequence of large amounts of the cementite phase, white iron is extremely hard but also very brittle, to the point of being virtually unmachinable. Its use is limited to applications that necessitate a very hard and wear-resistant surface, without a high degree of ductility—for example, as rollers in rolling mills.

Heating white iron at temperatures between 800 and 900°C for a prolonged time period and in a neutral atmosphere (to prevent oxidation) causes a decomposition of the cementite, forming graphite, which exists in the form of clusters or rosettes surrounded by a ferrite or pearlite matrix. Another cast iron, **malleable iron**, is produced. Representative applications include connecting rods, transmission gears, and differential cases for the automotive industry, and also flanges, pipe fittings, and valve parts for railroad, marine, and other heavy-duty services.

Gray and nodular cast irons are produced in approximately the same amounts; however, white and malleable cast irons are produced in smaller quantities.

Copper and its alloys

Copper and copper-based alloys, possessing a desirable combination of physical properties, have been utilized in quite a variety of applications since antiquity. The most common copper alloys are the **brasses** for which zinc, as a substitutional impurity, is the predominant alloying element. Some of the common uses for brass alloys include costume jewelry, cartridge casings, automotive radiators, musical instruments, electronic packaging, and coins. The **bronzes** are alloys of copper and several other elements, including tin, aluminum, silicon, and nickel. These alloys are somewhat stronger than the brasses, yet they still have a high degree of corrosion resistance. Generally they are utilized when, in addition to corrosion resistance, good tensile properties are required.

Aluminum and Its Alloys

Aluminum and its alloys are characterized by a relatively low density (2.7 g/cm^3 as compared to 7.9 g/cm^3 for steel), high electrical and thermal conductivities, and a resistance to corrosion in some common environments, including the ambient atmosphere. Many of these alloys are easily formed by virtue of high ductility; this is evidenced by the thin aluminum foil sheet into which the relatively pure material may be rolled.

Generally, aluminum alloys are classified as either cast or wrought. Some of the common applications of aluminum alloys include aircraft structural parts, beverage cans, bus bodies, and automotive parts (engine blocks, pistons, and manifolds). Recent attention has been given to alloys of aluminum and other low-density metals (e.g., Mg and Ti) as engineering materials for transportation, to effect reductions in fuel consumption. An important characteristic of these materials is specific strength, which is quantified by the tensile strength-specific gravity ratio. Even though an alloy of one of these metals may have a tensile strength that is inferior to a more dense material (such as steel), on a weight basis it will be able to sustain a larger load.

Magnesium and Its Alloys

Perhaps the most outstanding characteristic of magnesium is its density, 1.7 g/cm^3, which is the lowest of all the structural metals; therefore, its alloys are used where light weight is an important consideration (e.g., in aircraft components). At room temperature magnesium and its alloys are difficult to deform. Consequently, most fabrication is by casting or hot working at temperatures between 200 and 350 ℃. These alloys are used in aircraft and missile applications, as well as in luggage. Furthermore, in the last several years the demand for magnesium alloys has increased dramatically in a host of different industries. For many applications, magnesium alloys have replaced engineering plastics that have comparable densities inasmuch as the magnesium materials are stiffer, more recyclable, and less costly to produce. For example, magnesium is now employed in a variety of hand-held devices (e.g., chain saws, power tools, hedge clippers), in automobiles (e.g., steering wheels and columns, seat frames, transmission cases), and in audio-video-computer-communications equipment (e. g., laptop computers, camcorders, TV sets, cellular telephones).

Titanium and Its Alloys

Titanium and its alloys are relatively new engineering materials that possess an extraordinary combination of properties. Titanium alloys are extremely strong; room temperature tensile strengths as high as 1400 MPa are attainable, yielding remarkable specific strengths. Furthermore, the alloys are highly ductile and easily forged and machined. The corrosion resistance of titanium alloys at normal temperatures is unusually high; they are virtually immune to air, marine, and a variety of industrial environments. These alloys are commonly utilized in airplane structures, space vehicles, surgical implants, and in the petroleum and chemical industries.

The Refractory Metals

Metals that have extremely high melting temperatures are classified as the refractory metals. Included in this group are niobium (Nb), molybdenum (Mo), tungsten (W), and tantalum (Ta). Melting temperatures range between 2468℃, for niobium and 3410℃, the highest melting temperature of any metal, for tungsten. The applications of these metals are varied. For example, tantalum and molybdenum are alloyed with stainless steel to improve its corrosion resistance. Molybdenum alloys are utilized for extrusion dies and structural parts in space vehicles; incandescent light filaments, X-ray tubes, and welding electrodes employ tungsten alloys. Tantalum is immune to chemical attack by virtually all environments at temperatures below 150℃, and is frequently used in applications requiring such a corrosion-resistant material.

The Superalloys

The superalloys have superlative combinations of properties. Most are used in aircraft turbine components, which must withstand exposure to severely oxidizing environments and high temperatures for reasonable time periods. Mechanical integrity under these conditions is critical; in this regard, density is an important consideration because centrifugal stresses are diminished in rotating members when the density is reduced. These materials are classified according to the predominant metal in the alloy, which may be cobalt, nickel, or iron. Other alloying elements include the refractory metals (Nb, Mo, W, Ta), chromium, and titanium. In addition to turbine applications, these alloys are utilized in nuclear reactors and petrochemical equipment.

The Noble Metals

The noble or precious metals are a group of eight elements that have some physical characteristics in common. They are expensive (precious) and are superior or notable (noble) in properties—that is, characteristically soft, ductile, and oxidation resistant. The noble metals are silver, gold, platinum, palladium, rhodium, ruthenium, iridium, and osmium; the first three are most common and are used extensively in jewelry. Silver and gold may be strengthened by solid-solution alloying with copper; sterling silver is a silver-copper alloy containing approximately 7.5 wt% Cu. Alloys of both silver and gold are employed as dental restoration materials; also, some integrated circuit electrical contacts are of gold. Platinum is used for chemical laboratory equipment, as a catalyst (especially in the manufacture of gasoline), and in thermocouples to measure elevated temperatures.

Miscellaneous Nonferrous Alloys

The discussion above covers the vast majority of nonferrous alloys; however, a number of others are found in a variety of engineering applications, and a brief exposure of these is worthwhile.

Nickel and its alloys are highly resistant to corrosion in many environments, especially those that are basic (alkaline). Nickel is often coated or plated on some metals that are susceptible to corrosion as a protective measure. Monel, a nickel-based alloy containing approximately 65 wt% Ni and 28 wt% Cu (the balance iron), has very high strength and is extremely corrosion resistant; it is used in pumps, valves, and other components that are in contact with some acid and petroleum solutions.

Lead, tin, and their alloys find some use as engineering materials. Both are mechanically soft and weak, have low melting temperatures, are quite resistant to many corrosion environments, and have recrystallization temperatures below room temperature. Some common solders are lead-tin alloys, which have low melting temperatures. Applications for lead and its alloys include x-ray shields and storage batteries. The primary use of tin is as a very thin coating on the inside of plain carbon steel cans (tin cans) that are used for food containers; this coating inhibits chemical reactions between the steel and the food products.

Unalloyed zinc also is a relatively soft metal having a low melting temperature and a subambient recrystallization temperature. Chemically, it is reactive in a number of common environments and, therefore, susceptible to corrosion. Galvanized steel is just plain carbon steel that has been coated with a thin zinc layer; the zinc preferentially corrodes and protects the steel. Typical applications of galvanized steel are familiar (sheet metal, fences, screen, screws, etc.). Common applications of zinc alloys include padlocks, plumbing fixtures, automotive parts (door handles and grilles), and office equipment.

Zirconium and its alloys are ductile and have other mechanical characteristics that are comparable to those of titanium alloys and the austenitic stainless steels. However, the primary asset of these alloys is their resistance to corrosion in a host of corrosive media, including superheated water. Furthermore, zirconium is transparent to thermal neutrons, so that its alloys have been used as cladding for uranium fuel in water-cooled nuclear reactors. In terms of cost, these alloys are also often the materials of choice for heat exchangers, reactor vessels, and piping systems for the chemical processing and nuclear industries. They are also used in incendiary ordnance and in sealing devices for vacuum tubes.

4.5.3 Fabrication of Metals

Metal fabrication techniques are normally preceded by refining, alloying, and often heat-treating processes that produce alloys with the desired characteristics. The classifications of fabrication techniques include various metal-forming methods, casting, powder metallurgy, welding, and machining; often two or more of them must be used before a piece is finished. The methods chosen depend on several factors; the most important are the properties of the metal, the size and shape of the finished piece, and, of course, cost.

Forming operations

Forming operations are those in which the shape of a metal piece is changed by plastic deformation; for example, forging, rolling, extrusion, and drawing are common forming techniques. Of course, the deformation must be induced by an external force or stress, the magnitude of which must exceed the yield strength of the material. Most metallic materials are especially amenable to these procedures, being at least moderately ductile and capable of some permanent deformation without cracking or fracturing.

When deformation is achieved at a temperature above that at which recrystallization occurs, the process is termed **hot working**; otherwise, it is **cold working**. For hot-working operations, large deformations are possible, which may be successively repeated because the metal remains soft and ductile. Also, deformation energy requirements are less than for cold working. However, most metals experience some surface oxidation, which results in material loss and a poor final surface finish. Cold working produces an increase in strength with the attendant decrease in ductility, since the metal strain hardens; advantages over hot working include a higher quality surface finish, better mechanical properties and a greater variety of them, and closer dimensional control of the finished piece.

Casting

Casting is a fabrication process whereby a totally molten metal is poured into a mold cavity having the desired shape; upon solidification, the metal assumes the shape of the mold but experiences some shrinkage. Casting techniques are employed when ①the finished shape is so large or complicated that any other method would be impractical, ②a particular alloy is so low in ductility that forming by either hot or cold working would be difficult, and ③in comparison to other fabrication processes, casting is the most economical. A number of different casting techniques are commonly employed, including sand, die, investment, lost foam, and continuous casting.

Powder Metallurgy

This fabrication technique involves the compaction of powdered metal, followed by a heat treatment to produce a more dense piece. The process is appropriately called powder metallurgy, frequently designated as P/M. Powder metallurgy makes it possible to produce a virtually nonporous piece having properties almost equivalent to the fully dense parent material. Diffusional processes during the heat treatment are central to the development of these properties. This method is especially suitable for metals having low ductilities, since only small plastic deformation of the powder particles need occur. Metals having high melting temperatures are difficult to melt and cast, and fabrication is expedited using P/M. Furthermore, parts that require very close dimensional tolerances (e. g., bushings and gears) may be economically produced using this technique.

Welding

In welding, two or more metal parts are joined to form a single piece when one-part fabrication is expensive or inconvenient. Both similar and dissimilar metals may be welded. The joining bond is metallurgical (involving some diffusion) rather than just mechanical, as with riveting and bolting. A variety of welding methods exist, including arc and gas welding, as well as brazing and soldering.

4.5.4 Thermal Processing of Metals

A number of phenomena, which occur in metals and alloys at elevated temperatures——for example, recrystallization and the decomposition of austenite, are effective in altering the mechanical characteristics when appropriate heat treatments or thermal processes are employed. In fact, the use of heat treatments on commercial alloys is an exceedingly common practice.

Annealing processes

The term **annealing** refers to a heat treatment in which a material is exposed to an elevated temperature for an extended time period and then slowly cooled. Ordinarily, annealing is carried out to ①relieve stresses; ②increase softness, ductility, and toughness; and/or ③produce a specific microstructure. A variety of annealing heat treatments are possible; they are characterized by the changes that are induced, which many times are microstructural and are responsible for the alteration of the mechanical properties.

Heat treatment of steels

Conventional heat treatment procedures for producing martensitic steels ordinarily involve continuous and rapid cooling of an austenitized specimen in some type of quenching medium, such as water, oil, or air. The optimum properties of a steel that has been quenched and then tempered can be realized only if, during the quenching heat treatment, the specimen has been converted to a high content of martensite; the formation of any pearlite and/or bainite will result in other than the best combination of mechanical characteristics.

Precipitation hardening

The strength and hardness of some metal alloys may be enhanced by the formation of extremely small uniformly dispersed particles of a second phase within the original phase matrix; this must be accomplished by phase transformations that are induced by appropriate heat treatments. The process is called **precipitation hardening** because the small particles of the new phase are termed "precipitates." "Age hardening" is also used to designate this procedure because the strength develops with time, or as the alloy ages. Examples of alloys that are hardened by precipitation treatments include aluminum-copper, copper-beryllium, copper-tin, and magnesium-aluminum; some ferrous alloys are also precipitation hardenable.

Summary

Concepts of Stress and Strain
Stress-Strain Behavior
Tensile Properties

A number of the important mechanical properties of materials, predominantly metals, have been discussed in this chapter. Concepts of stress and strain were first introduced. Stress is a measure of an applied mechanical load or force, normalized to take into account cross-sectional area. Strain represents the amount of deformation induced by a stress.

Some of the mechanical characteristics of metals can be ascertained by simple stress-strain tests. A material that is stressed first undergoes elastic, or nonpermanent, deformation, wherein stress and strain are proportional. The constant of proportionality is the modulus of elasticity for tension and compression, and is the shear modulus when the stress is shear. Poisson's ratio represents the negative ratio of transverse and longitudinal strains.

The phenomenon of yielding occurs at the onset of plastic or permanent deformation; yield strength is determined by a strain offset method from the stress-strain behavior, which is indicative of the stress at which plastic deformation begins. Tensile strength corresponds to the maximum tensile stress that may be sustained by a specimen, whereas percents elongation and reduction in area are measures of ductility—the amount of plastic deformation that has occurred at fracture.

Hardness

Hardness is a measure of the resistance to localized plastic deformation. In several popular hardness-testing techniques (Rockwell, Brinell, Knoop, and Vickers) a small indenter is forced into the surface of the material, and an index number is determined on the basis of the size or depth of the resulting indentation. For many metals, hardness and tensile strength are approximately proportional to each other.

Slip Systems

On a microscopic level, plastic deformation corresponds to the motion of dislocations in response to an externally applied shear stress, and a process is termed "slip". Slip occurs on specific crystallographic planes and within these planes only in certain directions. A slip system represents a slip plane-slip direction combination, and operable slip systems depend on the crystal structure of the material.

Slip in Single Crystals

The critical resolved shear stress is the minimum shear stress required to initiate dislocation motion; the yield strength of a single crystal depends on both the magnitude of the critical resolved shear stress and the orientation of slip components relative to the direction of the applied stress.

Plastic Deformation of Polycrystalline Materials

For polycrystalline materials, slip occurs within each grain along the slip systems that are most favorably oriented with the applied stress; furthermore, during deformation, grains change shape in such a manner that coherency at the grain boundaries is maintained.

Characteristics of Dislocations

Strengthening by Grain Size Reduction

Solid-Solution Strengthening

Strain Hardening

Since the ease with which a material is capable of plastic deformation is a function of dislocation mobility, restricting dislocation motion increases hardness and strength. On the basis of this principle, three different strengthening mechanisms were discussed.

Grain boundaries serve as barriers to dislocation motion; thus refining the grain size of a polycrystalline material renders it harder and stronger. Solid-solution strengthening results from lattice strain interactions between impurity atoms and dislocations. Finally, as a material is plastically deformed, the dislocation density increases, as does also the extent of repulsive dislocation-dislocation strain field interactions; strain hardening is just the enhancement of strength with increased plastic deformation.

Recovery

Recrystallization

Grain Growth

The microstructural and mechanical characteristics of a plastically deformed metal specimen may be restored to their predeformed states by an appropriate heat treatment, during which recovery, recrystallization, and grain growth processes are allowed to occur. During recovery there is a reduction in dislocation density and alterations in dislocation configurations. Recrystallization is the formation of a new set of grains that are strain free; in addition, the material becomes softer and more ductile. Grain growth is the increase in average grain size of polycrystalline materials, which proceeds by grain boundary motion.

Failure

Ductile Fracture

Fracture, in response to tensile loading and at relatively low temperatures, may occur by ductile and brittle modes, both of which involve the formation and propagation of cracks. For ductile fracture, evidence will exist of gross plastic deformation at the fracture surface. In tension, highly ductile metals will neck down to essentially a point fracture; cup-and-cone mating fracture surfaces result for moderate ductility.

Cracks in ductile materials are said to be stable (i.e., resist extension without an increase in applied stress); and inasmuch as fracture is noncatastrophic, this fracture mode is almost always preferred.

Brittle Fracture

For brittle fracture, cracks are unstable, and the fracture surface is relatively flat and perpendicular to the direction of the applied tensile load. Chevron and ridgelike patterns are possible, which indicate the direction of crack propagation. Transgranular (through-grain) and intergranular (between-grain) fractures are found in brittle polycrystalline materials.

Principles of Fracture Mechanics

The significant discrepancy between actual and theoretical fracture strengths of brittle materials is explained by the existence of small flaws that are capable of amplifying an applied tensile stress in their vicinity, leading ultimately to crack formation. Fracture ensues when the theoretical cohesive strength is exceeded at the tip of one of these flaws. The fracture toughness of a material is indicative of its resistance to brittle fracture when a crack is present. It depends on specimen thickness, and, for relatively thick specimens (i.e., conditions of plane strain), is termed the plane strain fracture toughness.

Fatigue

Fatigue is a common type of catastrophic failure wherein the applied stress level fluctuates with time. Test data are plotted as stress versus the logarithm of the number of cycles to failure. For many metals and alloys, stress diminishes continuously with increasing number of cycles at failure; fatigue strength and fatigue life are the parameters used to characterize the fatigue behavior of these materials. On the other hand, for other metals/alloys, at some point, stress ceases to decrease with, and becomes independent of, the number of cycles; the fatigue behavior of these materials is expressed in terms of fatigue limit.

Creep

The time-dependent plastic deformation of materials subjected to a constant load (or stress) and the temperature greater than about $0.4T_m$ is termed creep. A typical creep curve (strain versus time) will normally exhibit three distinct regions. For transient (or primary) creep, the rate (or slope) diminishes with time. The plot becomes linear (i.e., creep rate is constant) in the steady-state (or secondary) region. And finally, deformation accelerates for tertiary creep, just prior to failure (or rupture). Important design parameters available from such a plot include the steady-state creep rate (slope of the linear region) and rupture lifetime.

Phase Equilibria

One-Component (or Unary) Phase Diagrams

Binary Phase Diagrams

Interpretation of Phase Diagrams

Equilibrium phase diagrams are a convenient and concise way of representing the most stable relationships between phases in alloy systems. This discussion began by considering the unary (or pressure-temperature) phase diagram for a one component system. Solid-, liquid-, and vapor-phase regions are found on this type of phase diagram. For binary systems, temperature and composition are variables, whereas external pressure is held constant. Areas, or phase regions, are defined on these temperature-versus-composition plots within which either one or two phases exist. For an alloy of specified composition and at a known temperature, the phases present, their compositions, and relative amounts under equilibrium conditions may be determined. Within two-phase regions, tie lines and the lever rule must be used for phase composition and mass fraction computations, respectively.

Binary Isomorphous Systems

Development of Microstructure in Isomorphous Alloys

Several different kinds of phase diagram were discussed for metallic systems. Isomorphous diagrams are those for which there is complete solubility in the solid phase; the copper-nickel system displays this behavior. Also discussed for alloys belonging to isomorphous systems were the development of microstructure for both cases of equilibrium and nonequilibrium cooling, and the dependence of mechanical characteristics on composition.

Binary Eutectic Systems

Development of Microstructure in Eutectic Alloys

In a eutectic reaction, as found in some alloy systems, a liquid phase transforms isothermally to two different solid phases upon cooling. Such a reaction is noted on the copper-silver and lead-tin phase diagrams. Complete solid solubility for all compositions does not exist; instead, solid solutions are terminal-there is only a limited solubility of each component in the other. Four different kinds of microstructures that may develop for the equilibrium cooling of alloys belonging to eutectic systems were discussed.

The Iron-Iron Carbide (Fe-Fe_3C) Phase Diagram
Development of Microstructure in Iron-Carbon Alloys

Considerable attention was given to the iron-carbon system, and specifically, the iron-iron carbide phase diagram, which technologically is one of the most important. The development of microstructure in many iron-carbon alloys and steels depends on the eutectoid reaction in which the FCC austenite phase of composition 0.76 wt% C transforms isothermally to the BCC ferrite phase (0.022 wt% C) and the intermetallic compound, cementite (Fe_3C). The microstructural product of an iron-carbon alloy of eutectoid composition is pearlite, a microconstituent consisting of alternating layers of ferrite and cementite. The microstructures of alloys having carbon contents less than the eutectoid (hypoeutectoid) are comprised of a proeutectoid ferrite phase in addition to pearlite. On the other hand, pearlite and proeutectoid cementite constitute the microconstituents for hypereutectoid alloys—those with carbon contents in excess of the eutectoid composition.

Ferrous Alloys

This chapter began with a discussion of the various types of metal alloys. With regard to composition, metals and alloys are classified as either ferrous or nonferrous.

Ferrous alloys (steels and cast irons) are those in which iron is the prime constituent.

Most steels contain less than 1.0 wt% C, and, in addition, other alloying elements, which render them susceptible to heat treatment (and an enhancement of mechanical properties) and/or more corrosion resistant. Plain low-carbon steels and high-strength low-alloy, medium-carbon, tool, and stainless steels are the most common types.

Cast irons contain a higher carbon content, normally between 3.0 and 4.5 wt%C, and other alloying elements, notably silicon. For these materials, most of the carbon exists in graphite form rather than combined with iron as cementite. Gray, ductile (or nodular), malleable, and compacted graphite irons are the four most widely used cast irons; the latter three are reasonably ductile.

Nonferrous Alloys

All other alloys fall within the nonferrous category, which is further subdivided according to base metal or some distinctive characteristic that is shared by a group of alloys. The compositions, typical properties, and applications of copper, aluminum, magnesium, titanium, nickel, lead, tin, zirconium, and zinc alloys, as well as the refractory metals, the superalloys, and the noble metals were discussed.

Fabrication of metals

Metal fabrication techniques are normally preceded by refining, alloying, and often heat-treating processes that produce alloys with the desired characteristics. The classifications of fabrication techniques include various metal-forming methods, casting, powder metallurgy, welding, and machining; often two or more of them must be used before a piece is finished. The methods chosen depend on several factors; the most important are the properties of the metal, the size and shape of the finished piece, and, of course, cost.

Important Terms and Concepts

Modulus of elasticity	Shear modulus	Ductility
Elastic deformation	Proportional limit	Fracture toughness
Plastic deformation	Hardness	Tensile strength
Engineering stress	Poisson's ratio	Yield strength
Engineering strain	True stress	True strain
Cold working	Critical shear stress	Dislocation density
Recovery	Recrystallization	Grain growth
Recrystallization temperature	Strain hardening	Solid-solution strengthening
Brittle fracture	Ductile fracture	Fracture toughness
Corrosion fatigue	Fatigue strength	Fracture mechanics
Ductile-to-brittle transition	Impact energy	Intergranular fracture
Fatigue	Creep	Transgranular fracture
Fatigue limit	Thermal fatigue	Eutectic phase
Fatigue life	Cementite	Eutectoid reaction
Stress raiser	Equilibrium	Gibbs phase rule
Austenite	Eutectic structure	Intermediate solid solution
Eutectic reaction	Free energy	Isomorphous
Ferrite	Hypoeutectoid alloy	Metastable
Hypereutectoid alloy	Invariant point	Peritectic reaction
Intermetallic compound	Liquidus line	Phase equilibrium
Lever rule	Pearlite	Proeutectoid ferrite
Tie line	Proeutectoid cementite	Artificial aging
Primary phase	Solubility limit	Natural aging
Solidus line	Terminal solid solution	Compacted graphite iron
Solvus line	Ductile iron	Extrusion
Alloy steel	Bronze	Hardenability
Ferrous alloy	Process annealing	White cast iron
Brass	Welding	Cast iron
Solution heat treatment	Gray cast iron	Powder metallurgy(P/M)
Stainless steel	Normalizing	Precipitation hardening
Plain carbon steel	Full annealing	Nonferrous alloy
Forging	Annealing	Hot working
Drawing		Cold working
Austenitizing		Tempering

References

[1] *ASM Handbook*, Vol. 8, *Mechanical Testing and Evaluation*, ASM International, Materials Park, OH, 2000.

[2] Boyer, H. E. (Editor), *Atlas of Stress-Strain Curves*, 2nd edition, ASM International, Materials Park, OH, 2002.

[3] Chandler, H. (Editor), *Hardness Testing*, 2nd edition, ASM International, Materials Park, OH, 2000.

[4] Courtney, T. H., *Mechanical Behavior of Materials*, 2nd edition, McGraw-Hill Higher Education, Burr Ridge, IL, 2000.

[5] Davis, J. R. (Editor), *Tensile Testing*, 2nd edition, ASM International, Materials Park, OH, 2004.

[6] Dieter, G. E., *Mechanical Metallurgy*, 3rd edition, McGraw-Hill Book Company, New York, 1986.

[7] Dowling, N. E., *Mechanical Behavior of Materials*, 2nd edition, Prentice Hall PTR, Paramus, NJ, 1998.

[8] McClintock, F. A. and A. S. Argon, *Mechanical Behavior of Materials*, Addison-Wesley Publishing Co., Reading, MA, 1966. Reprinted by CBLS Publishers, Marietta, OH, 1993.

[9] Meyers, M. A. and K. K. Chawla, *Mechanical Behavior of Materials*, Prentice Hall PTR, Paramus, NJ, 1999.

[10] Hirth, J. P., and J. Lothe, *Theory of Dislocations*, 2nd edition, Wiley-Interscience, New York, 1982. Reprinted by Krieger, Melbourne, FL, 1992.

[11] Hull, D., *Introduction to Dislocations*, 3rd edition, Butterworth-Heinemann, Woburn, UK, 1984. Read, W. T., Jr., *Dislocations in Crystals*, McGraw-Hill, New York, 1953.

[12] Weertman, J., and J. R. Weertman, *Elementary Dislocation Theory*, Macmillan, New York, 1964. Reprinted by Oxford University Press, New York, 1992.

[13] *ASM Handbook*, Vol. 11, *Failure Analysis and Prevention*, ASM International, Materials Park, OH, 1986.

[14] *ASM Handbook*, Vol. 12, *Fractography*, ASM International, Materials Park, OH, 1987.

[15] *ASM Handbook*, Vol. 19, *Fatigue and Fracture*, ASM International, Materials Park, OH, 1996.

[16] Boyer, H. E. (Editor), *Atlas of Creep and Stress-Rupture Curves*, ASM International, Materials Park, OH, 1988.

[17] Boyer, H. E. (Editor), *Atlas of Fatigue Curves*, ASM International, Materials Park, OH, 1986.

[18] Colangelo, V. J. and F. A. Heiser, *Analysis of Metallurgical Failures*, 2nd edition, Wiley, New York, 1987.

[19] Collins, J. A., *Failure of Materials in Mechanical Design*, 2nd edition, Wiley, New York, 1993. Courtney, T. H., *Mechanical Behavior of Materials*, 2nd edition, McGraw-Hill, New York, 2000.

[20] Dieter, G. E., *Mechanical Metallurgy*, 3rd edition, McGraw-Hill, New York, 1986.

[21] Esaklul, K. A., *Handbook of Case Histories in Failure Analysis*, ASM International, Materials Park, OH, 1992 and 1993. In two volumes.

[22] *Fatigue Data Book: Light Structural Alloys*, ASM International, Materials Park, OH, 1995.

[23] Hertzberg, R. W., *Deformation and Fracture Mechanics of Engineering Materials*, 4th edition, Wiley, New York, 1996.

[24] Stevens, R. I., A. Fatemi, R. R. Stevens, and H. O. Fuchs, *Metal Fatigue in Engineering*, 2nd edition, Wiley, New York, 2000.

[25] Tetelman, A. S. and A. J. McEvily, *Fracture of Structural Materials*, Wiley, New York, 1967. Reprinted by Books on Demand, Ann Arbor, MI.

[26] Wulpi, D. J., *Understanding How Components Fail*, 2nd edition, ASM International, Materials Park, OH, 1999.

[27] *ASM Handbook*, Vol. 3, *Alloy Phase Diagrams*, ASM International, Materials Park, OH, 1992.

[28] *ASM Handbook*, Vol. 9, *Metallography and Microstructures*, ASM International, Materials Park, OH, 2004.

[29]   Hansen, M. and K. Anderko, *Constitution of Binary Alloys*, 2nd edition, McGraw-Hill, New York, 1958. *First Supplement* (R. P. Elliott), 1965. *Second Supplement* (F. A. Shunk), 1969. Reprinted by Genium Publishing Corp., Schenectady, NY.

[30]   T. B. Massalski, H. Okamoto, P. R. Subramanian, and L. Kacprzak (Editors), *Binary Phase Diagrams*, 2nd edition, ASM International, Materials Park, OH, 1990. Three volumes.

[31]   Okamoto, H., *Desk Handbook: Phase Diagrams for Binary Alloys*, ASM International, Materials Park, OH, 2000.

[32]   Atkins, M., *Atlas of Continuous Cooling Transformation Diagrams for Engineering Steels*, British Steel Corporation, Sheffield, England, 1980.

[33]   *Atlas of Isothermal Transformation and Cooling Transformation Diagrams*, ASM International, Materials Park, OH, 1977.

[34]   Brooks, C. R., *Principles of the Heat Treatment of Plain Carbon and Low Alloy Steels*, ASM International, Materials Park, OH, 1996.

[35]   Porter, D. A. and K. E. Easterling, *Phase Transformations in Metals and Alloys*, Chapman and Hall, New York, 1992.

[36]   Shewmon, P. G., *Transformations in Metals*, McGraw-Hill, New York, 1969. Reprinted by Williams Book Company, Tulsa, OK.

[37]   Vander Voort, G. (Editor), *Atlas of Time-Temperature Diagrams for Irons and Steels*, ASM International, Materials Park, OH, 1991.

[38]   Brooks, C. R., *Principles of the Heat Treatment of Plain Carbon and Low Alloy Steels*, ASM International, Materials Park, OH, 1995.

Chapter 5 Ceramic Materials

Learning Objectives

After careful study of this chapter you should be able to do the following:
1. Sketch/describe unit cells for sodium chloride, cesium chloride, zinc blende, rock salt, silicate ceramics and carbon crystal structures.
2. Briefly explain why there is normally significant scatter in the fracture strength for identical specimens of the same ceramic material.
3. Compute the flexural strength of ceramic rod specimens that have been bent to fracture in three-point loading.
4. Name the two types of clay products, and then give two examples of each.
5. Cite three important requirements that normally must be met by refractory ceramics and abrasive ceramics.
6. Describe the process that is used to produce glasses and glass-ceramics.
7. Briefly describe processes that occur during the drying and firing of clay-based ceramic ware.

5.1 Structures and Properties of Ceramics

5.1.1 Introduction

The utilization of ceramic materials by man is probably as old as human civilization itself. Stone, obsidian, clay, quartz, and mineral ores are as much a part of the history of mankind as the products which have been made of them. Among these products are tools, earthenware, stoneware, porcelain, as well as bricks, refractories, body paints, insulators, abrasives, and eventually modern "high-tech ceramics" used, for example, in electronic equipment or jet engines. Actually, fired or baked ceramic objects are probably the oldest existing samples of handicraft which have come to us from ancient times. They are often the only archaeological clues that witness former civilizations and habitats. Taking all of these components into consideration, it might be well justified to ask why historians did not specifically designate a ceramics age. The answer is quite simple: stone, copper, bronze, and iron can be associated with reasonably well-defined time periods that have a beginning and frequently also an end during which these materials were predominantly utilized for the creation of tools, weapons, and objects of art. In contrast to this, ceramic materials have been used continuously by man with essentially unbroken vigor commencing from many millennia ago until the present time.

5.1.2 Ceramic Structures

Because ceramics are composed of at least two elements, and often more, their crystal structures are generally more complex than those for metals. The atomic bonding in these materials ranges from purely ionic to totally covalent; many ceramics exhibit a combination of these two bonding types, the degree of ionic character being dependent on the electronegativities of the atoms.

Crystal structures

For those ceramic materials for which the atomic bonding is predominantly ionic, the crystal structures may be thought of as being composed of electrically charged ions instead of atoms. The metallic ions, or **cations**, are positively charged, because they have given up their valence electrons to the nonmetallic ions, or **anions**, which are negatively charged. Two characteristics of the component ions in crystalline ceramic materials influence the crystal structure: the magnitude of the electrical charge on each of the component ions, and the relative sizes of the cations and anions. With regard to the first characteristic, the crystal must be electrically neutral; that is, all the cation positive charges must be balanced by an equal number of anion negative charges. The second criterion involves the sizes or ionic radii of the cations and anions. Because the metallic elements give up electrons when ionized, cations are ordinarily smaller than anions. Each cation prefers to have as many nearest-neighbor anions as possible. The anions also desire a maximum number of cation nearest neighbors.

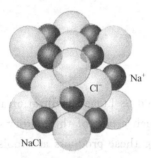

Figure 5.1 A unit cell for the rock salt, or sodium chloride (NaCl), crystal structure

Some of the common ceramic materials are those in which there are equal numbers of cations and anions. These are often referred to as AX compounds, where A denotes the cation and X the anion. There are several different crystal structures for AX compounds; each is normally named after a common material that assumes the particular structure. Perhaps the most common AX crystal structure is the sodium chloride (NaCl), or *rock salt*, type. The coordination number for both cations and anions is 6, and therefore the cation-anion radius ratio is between approximately 0.414 and 0.732. A unit cell for this crystal structure (Figure 5.1) is generated from an FCC arrangement of anions with one cation situated at the cube center and one at the center of each of the 12 cube edges. Some of the common ceramic materials that form with this crystal structure are NaCl, MgO, MnS, LiF, and FeO.

Figure 5.2 shows a unit cell for the cesium chloride (CsCl) crystal structure; the coordination number is 8 for both ion types. The anions are located at each of the corners of a cube, whereas the cube center is a single cation. Interchange of anions with cations, and vice versa, produces the same crystal structure. This is not a BCC crystal structure because ions of two different kinds are involved.

A third AX structure is one in which the coordination number is 4; that is, all ions are tetrahedrally coordinated. This is called the *zinc blende*, or *sphalerite*, structure, after the mineralogical term for zinc sulfide (ZnS). A unit cell is presented in Figure 5.3.

If the charges on the cations and anions are not the same, a compound can exist with the chemical formula $A_m X_p$, where m and/or $p \neq 1$. An example would be AX_2, for which a common crystal structure is found in fluorite (CaF_2).

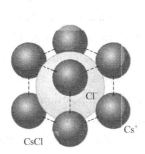

Figure 5.2 A unit cell for the cesium chloride (CsCl) crystal structure

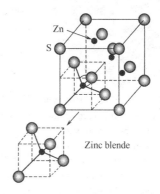

Figure 5.3 A unit cell for the zinc blende (ZnS) crystal structure

It is also possible for ceramic compounds to have more than one type of cation; for two types of cations (represented by A and B), their chemical formula may be designated as $A_m B_n X_p$. Barium titanate ($BaTiO_3$) having both Ba^{2+} and Ti^{4+} cations, falls into this classification. Of course, many other ceramic crystal structures are possible.

A number of ceramic crystal structures may be considered in terms of close-packed planes of ions, as well as unit cells. Ordinarily, the close-packed planes are composed of the large anions. As these planes are stacked atop each other, small interstitial sites are created between them in which the cations may reside.

These interstitial positions exist in two different types, as illustrated in Figure 5.4. Four atoms (three in one plane, and a single one in the adjacent plane) surround one type; this is termed a **tetrahedral position**. The other site type in Figure 5.4, involves six ion spheres, three in each of the two planes. Because an octahedron is produced by joining these six sphere centers, this site is called an **octahedral position**.

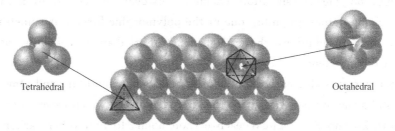

Figure 5.4 The stacking of one plane of close-packed (top) spheres (anions) on top of another (bottom); the geometries of tetrahedral and octahedral positions between the planes are noted

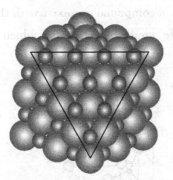

Figure 5.5 A section of the rock salt crystal structure from which a corner has been removed. The exposed plane of anions (big spheres inside the triangle) is a {111}-type plane; the cations (small spheres) occupy the interstitial octahedral positions

Ceramic crystal structures of this type depend on two factors: ① the stacking of the close-packed anion layers, and ② the manner in which the interstitial sites are filled with cations. For example, consider the rock salt crystal structure discussed above. The unit cell has cubic symmetry, and each cation has six ion nearest neighbors, as may be verified from Figure 5.1. That is, the ion at the center has as nearest neighbors the six ions that reside at the centers of each of the cube faces. The cations reside in octahedral positions because they have as nearest neighbors six anions. Furthermore, all octahedral positions are filled, since there is a single octahedral site per anion, and the ratio of anions to cations is 1 : 1. For this crystal structure, the relationship between the unit cell and close-packed anion plane stacking schemes is illustrated in Figure 5.5.

Silicate ceramics

Silicates are materials composed primarily of silicon and oxygen, the two most abundant elements in the earth's crust; consequently, the bulk of soils, rocks, clays, and sand come under the silicate classification. Rather than characterizing the crystal structures of these materials in terms of unit cells, it is more convenient to use various arrangements of a SiO_4^{4-} tetrahedron.

Often the silicates are not considered to be ionic because there is a significant covalent character to the interatomic Si-O bonds, which are directional and relatively strong. Regardless of the character of the Si-O bond, there is a -4 charge associated with every SiO_4^{4-} tetrahedron, since each of the four oxygen atoms requires an extra electron to achieve a stable electronic structure. Various silicate structures arise from the different ways in which the SiO_4^{4-} units can be combined into one-, two-, and three-dimensional arrangements.

Carbon

Carbon is an element that exists in various polymorphic forms, as well as in the amorphous state. This group of materials does not really fall within any one of the traditional metal, ceramic, polymer classification schemes. However, we choose to discuss these materials in this chapter since graphite, one of the polymorphic forms, is sometimes classified as a ceramic, and, in addition, the crystal structure of diamond, another polymorph, is similar to that of zinc blende.

Diamond is a metastable carbon polymorph at room temperature and atmospheric pressure. Its crystal structure is a variant of the zinc blende, in which carbon atoms occupy all positions (both Zn and S). Thus, each carbon bonds to four other carbons, and these bonds are totally covalent. This is appropriately called the diamond cubic crystal structure. The physical properties of diamond make it an extremely attractive material. It is extremely hard (the hardest known material) and has a very low electrical conductivity; these characteristics are due to its crystal structure and the strong interatomic covalent bonds. Furthermore, it has an unusually high thermal conductivity for a nonmetallic material, is optically transparent in the visible and infrared regions of the electromagnetic spectrum, and has a high index of refraction. Relatively large diamond single crystals are used as gem stones. Industrially, diamonds are utilized to grind or cut other softer materials.

Another polymorph of carbon is graphite; it has a crystal structure distinctly different from that of diamond and is also more stable than diamond at ambient temperature and pressure. The graphite structure is composed of layers of hexagonally arranged carbon atoms; within the layers, each carbon atom is bonded to three coplanar neighbor atoms by strong covalent bonds. The fourth bonding electron participates in a weak van der Waals type of bond between the layers. As a consequence of these weak interplanar bonds, interplanar cleavage is facile, which gives rise to the excellent lubricative properties of graphite. Also, the electrical conductivity is relatively high in crystallographic directions parallel to the hexagonal sheets.

Another polymorphic form of carbon was discovered in 1985. It exists in discrete molecular form and consists of a hollow spherical cluster of sixty carbon atoms; a single molecule is denoted by C_{60}. Each molecule is composed of groups of carbon atoms that are bonded to one another to form both hexagon (six-carbon atom) and pentagon (five-carbon atom) geometrical configurations. One such molecule, shown in Figure 5.6, is found to consist of 20 hexagons and 12 pentagons, which are arrayed such that no two pentagons share a common side; the molecular surface thus exhibits the symmetry of a soccer ball. The material com-

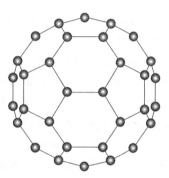

Figure 5.6 The structure of a C_{60} molecule

posed of C_{60} molecules is known as *buckminsterfullerene*, named in honor of R. Buckminster Fuller, who invented the geodesic dome; each C_{60} is simply a molecular replica of such a dome, which is often referred to as "buckyball" for short. The term **fullerene** is used to denote the class of materials that are composed of this type of molecule.

5.1.3 Mechanical Properties of Ceramics
Fracture of ceramics

Ceramic materials are somewhat limited in applicability by their mechanical properties, which in many respects are inferior to those of metals. The principal drawback is a disposition to catastrophic fracture in a brittle manner with very little energy absorption. At room temperature, both crystalline and noncrystalline ceramics almost always fracture before any plastic deformation can occur in response to an applied tensile load.

The brittle fracture process consists of the formation and propagation of cracks through the cross section of material in a direction perpendicular to the applied load. Crack growth in crystalline ceramics may be either transgranular (i.e., through the grains) or intergranular (i.e., along grain boundaries); for transgranular fracture, cracks propagate along specific crystallographic (or cleavage) planes, planes of high atomic density.

The measured fracture strengths of ceramic materials are substantially lower than predicted by theory from interatomic bonding forces. This may be explained by very small and omnipresent flaws in the material that serve as stress raisers-points at which the magnitude of an applied tensile stress is amplified. The degree of stress amplification depends on crack length and tip radius of curvature, being greatest for long and pointed flaws. These stress raisers may be minute surface or interior cracks (microcracks), internal pores, and grain corners, which are virtually impossible to eliminate or control. A stress concentration at a flaw tip can cause a crack to form, which may propagate until the eventual failure.

There is usually considerable variation and scatter in the fracture strength for many specimens of a specific brittle ceramic material. This phenomenon may be explained by the dependence of fracture strength on the probability of the existence of a flaw that is capable of initiating a crack. This probability varies from specimen to specimen of the same material and depends on fabrication technique and any subsequent treatment. Specimen size or volume also influences fracture strength; the larger the specimen, the greater is this flaw existence probability, and the lower the fracture strength.

For compressive stresses, there is no stress amplification associated with any existent flaws. For this reason, brittle ceramics display much higher strengths in compression than in tension (on the order of a factor of 10), and they are generally utilized when load conditions are compressive. Also, the fracture strength of a brittle ceramic may be enhanced dramatically by imposing residual compressive stresses at its surface. One way this may be accomplished is by thermal tempering.

Stress-strain behavior

The stress-strain behavior of brittle ceramics is not usually ascertained by a tensile test for three reasons. First, it is difficult to prepare and test specimens having the required geometry. Second, it is difficult to grip brittle materials without fracturing them; and third, ceramics fail after only about 0.1% strain, which necessitates that tensile specimens be perfectly aligned to avoid the presence of bending stresses, which are not easily calculated. Therefore, a more suitable transverse bending test is most frequently employed, in which a rod specimen having either a circular or rectangular cross section is bent until fracture using a three- or four-point loading technique; the three-point loading scheme is illustrated in Figure 5.7. Stress is computed from the specimen thickness, the bending moment, and the moment of inertia of the cross section; these parameters are noted in Figure 5.8 for rectangular and circular cross sections. The maximum tensile stress (as determined using these stress expressions) exists at the bottom specimen surface directly below the point of load application.

The stress at fracture using this flexure test is known as the **flexural strength**, *modulus of rupture*, *fracture strength*, or the *bend strength*, an important mechanical parameter for brittle ceramics. The elastic stress-strain behavior for ceramic materials using these flexure tests is similar to the tensile test results for metals: a linear relationship exists between stress and strain.

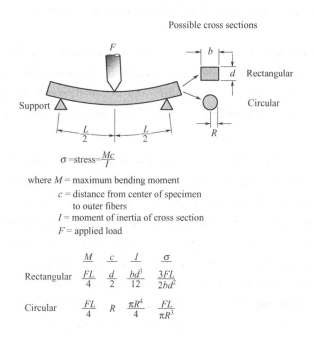

Figure 5.7 A three-point loading scheme for measuring the stress-strain behavior and flexural strength of brittle ceramics, including expressions for computing stress for rectangular and circular cross sections

5.2 Application and Processing of Ceramics

5.2.1 Types and Applications of Ceramics

Glasses

The glasses are a familiar group of ceramics; containers, lenses, and fiberglass represent typical applications. They are noncrystalline silicates containing other oxides, notably CaO, Na_2O, K_2O, and Al_2O_3, which influence the glass properties. A typical soda-lime glass consists of approximately 70 wt% SiO_2, the balance being mainly Na_2O (soda) and CaO (lime). Possibly the two prime assets of these materials are their optical transparency and the relative ease with which they may be fabricated.

Glass-ceramics

Most inorganic glasses can be made to transform from a noncrystalline state to one that is crystalline by the proper high-temperature heat treatment. This process is called crystallization, and the product is a fine-grained polycrystalline material which is often called a **glass-ceramic**. The formation of these small glass-ceramic grains is, in a sense, a phase transformation, which involves nucleation and growth stages. Glass-ceramic materials have been designed to have the following characteristics: relatively high mechanical strengths; low coefficients of thermal expansion (to avoid thermal shock); relatively high temperature capabilities; good dielectric properties (for electronic packaging applications); and good biological compatibility.

The most common uses for these materials are as ovenware, tableware, oven windows, and range tops—primarily because of their strength and excellent resistance to thermal shock. They also serve as electrical insulators and as substrates for printed circuit boards, and are used for architectural cladding, and for heat exchangers and regenerators.

Clay products

One of the most widely used ceramic raw materials is clay. This inexpensive ingredient, found naturally in great abundance, often is used as mined without any upgrading of quality. Another reason for its popularity lies in the ease with which clay products may be formed; when mixed in the proper proportions, clay and water form a plastic mass that is very amenable to shaping. The formed piece is dried to remove some of the moisture, after which it is fired at an elevated temperature to improve its mechanical strength.

Most of the clay-based products fall within two broad classifications: the **structural clay products** and the **whitewares**. Structural clay products include building bricks, tiles, and sewer pipes—applications in which structural integrity is important. The whiteware ceramics become white after the high-temperature firing. Included in this group are porcelain, pottery, tableware, china, and plumbing fixtures (sanitary ware).

Refractories

Another important class of ceramics that are utilized in large tonnages is the refractory ceramics. The salient properties of these materials include the capacity to withstand high temperatures without melting or decomposing, and the capacity to remain unreactive and inert when exposed to severe environments. In addition, the ability to provide thermal insulation is often an important consideration. Refractory materials are marketed in a variety of forms, but bricks are the most common. Typical applications include furnace linings for metal refining, glass manufacturing, metallurgical heat treatment, and power generation.

Abrasives

Abrasive ceramics are used to wear, grind, or cut away other material, which necessarily is softer. Therefore, the prime requisite for this group of materials is hardness or wear resistance; in addition, a high degree of toughness is essential to ensure that the abrasive particles do not easily fracture. Furthermore, high temperatures may be produced from abrasive frictional forces, so some refractoriness is also desirable.

Diamonds, both natural and synthetic, are utilized as abrasives; however, they are relatively expensive. The more common ceramic abrasives include silicon carbide, tungsten carbide (WC), aluminum oxide (or corundum), and silica sand.

Abrasives are used in several forms—bonded to grinding wheels, as coated abrasives, and as loose grains. Grinding, lapping, and polishing wheels often employ loose abrasive grains that are delivered in some type of oil- or water-based vehicle. Diamonds, corundum, silicon carbide, and rouge (an iron oxide) are used in loose form over a variety of grain size ranges.

Cements

Several familiar ceramic materials are classified as inorganic cements: cement, plaster of paris, and lime, which, as a group, are produced in extremely large quantities. The characteristic feature of these materials is that when mixed with water, they form a paste that subsequently sets and hardens. This trait is especially useful in that solid and rigid structures having just about any shape may be expeditiously formed. Also, some of these materials act as a bonding phase that chemically binds particulate aggregates into a single cohesive structure.

Of this group of materials, portland cement is consumed in the largest tonnages. It is produced by grinding and intimately mixing clay and lime-bearing minerals in the proper proportions, and then heating the mixture to about in a rotary kiln; this process, sometimes called calcination, produces physical and chemical changes in the raw materials. The resulting "clinker" product is then ground into a very fine powder to which is added a small amount of gypsum ($CaSO_4 \cdot 2H_2O$) to retard the setting process. This product is portland cement. The properties of portland cement, including setting time and final strength, to a large degree depend on its composition.

Advanced ceramics

Although the traditional ceramics discussed previously account for the bulk of the production, the development of new and what are termed "advanced ceramics" has begun and will continue to establish a prominent niche in our advanced technologies. In particular, electrical, magnetic, and optical properties and property combinations unique to ceramics have been exploited in a host of new products. Furthermore, advanced ceramics are utilized in optical fiber communications systems, as ball bearings, and in applications that exploit the piezoelectric behavior of a number of ceramic materials.

One new and advanced ceramic material that is a critical component in our modern optical communications systems is the optical fiber. The optical fiber is made of extremely high-purity silica, which must be free of even minute levels of contaminants and other defects that absorb, scatter, and attenuate a light beam. Very advanced and sophisticated processing techniques have been developed to produce fibers that meet the rigid restrictions required for this application.

Another new and interesting application of ceramic materials is in bearings. A bearing consists of balls and races that are in contact with and rub against one another when in use. In the past, both ball and race components traditionally have been made of bearing steels that are very hard, extremely corrosion resistant, and may be polished to a very smooth surface finish. Over the past decade or so silicon nitride (Si_3N_4) balls have begun replacing steel balls in a number of applications, since several properties of Si_3N_4 make it a more desirable material.

5.2.2 Fabrication and Processing of Ceramics

One chief concern in the application of ceramic materials is the method of fabrication.

Since ceramic materials have relatively high melting temperatures, casting them is normally impractical. Furthermore, in most instances the brittleness of these materials precludes deformation. Some ceramic pieces are formed from powders that must ultimately be dried and fired. Glass shapes are formed at elevated temperatures from a fluid mass that becomes very viscous upon cooling. Cements are shaped by placing into forms a fluid paste that hardens and assumes a permanent set by virtue of chemical reactions.

Fabrication and processing of glasses and glass-ceramics

Glass is produced by heating the raw materials to an elevated temperature above which melting occurs. Most commercial glasses are of the silica-soda-lime variety; the silica is usually supplied as common quartz sand, whereas Na_2O and CaO are added as soda ash (Na_2CO_3) and limestone ($CaCO_3$). For most applications, especially when optical transparency is important, it is essential that the glass product be homogeneous and pore free. Homogeneity is achieved by complete melting and mixing of the raw ingredients. Porosity results from small gas bubbles that are produced; these must be absorbed into the melt or otherwise eliminated, which requires proper adjustment of the viscosity of the molten material.

Four different forming methods are used to fabricate glass products: pressing, blowing, drawing, and fiber forming. Pressing is used in the fabrication of relatively thick-walled pieces such as plates and dishes. The glass piece is formed by pressure application in a graphite-coated cast iron mold having the desired shape; the mold is ordinarily heated to ensure an even surface.

The first stage in the fabrication of a glass-ceramic ware is forming it into the desired shape as a glass. Forming techniques used are the same as for glass pieces, i.e., pressing and drawing. Conversion of the glass into a glass-ceramic (i.e., crystallation) is accomplished by appropriate heat treatments. After melting and forming operations, nucleation and growth of the crystalline phase particles are carried out isothermally at two different temperatures.

Fabrication and processing of clay products

Clays are aluminosilicates, being composed of alumina (Al_2O_3) and silica (SiO_2), that contain chemically bound water. They have a broad range of physical characteristics, chemical compositions, and structures; common impurities include compounds (usually oxides) of barium, calcium, sodium, potassium, and iron, and also some organic matter. Crystal structures for the clay minerals are relatively complicated; however, one prevailing characteristic is a layered structure. The most common clay minerals that are of interest have what is called the kaolinite structure. When water is added, the water molecules fit in between these layered sheets and form a thin film around the clay particles. The particles are thus free to move over one another, which accounts for the resulting plasticity of the water-clay mixture.

The as-mined raw materials usually have to go through a milling or grinding operation in which particle size is reduced; this is followed by screening or sizing to yield a powdered product having a desired range of particle sizes. For multicomponent systems, powders must be thoroughly mixed with water and perhaps other ingredients to give flow characteristics that are compatible with the particular forming technique. The formed piece must have sufficient mechanical strength to remain intact during transporting, drying, and firing operations. Two common shaping techniques are utilized for forming clay-based compositions: hydroplastic forming and slip casting.

The most common hydroplastic forming technique is extrusion, in which a stiff plastic ceramic mass is forced through a die orifice having the desired cross-sectional geometry; it is similar to the extrusion of metals. Brick, pipe, ceramic blocks, and tiles are all commonly fabricated using hydroplastic forming. Usually the plastic ceramic is forced through the die by means of a motor-driven auger, and often air is removed in a vacuum chamber to enhance the density. Hollow internal columns in the extruded piece (e.g., building brick) are formed by inserts situated within the die.

Another forming process used for clay-based compositions is slip casting. A slip is a suspension of clay and/or other nonplastic materials in water. When poured into a porous mold (commonly made of plastic of paris), water from the slip is absorbed into the mold, leaving behind a solid layer on the mold wall the thickness of which depends on the time. As the cast piece dries and shrinks, it will pull away (or release) from the mold wall; at this time the mold may be disassembled and the cast piece removed.

A ceramic piece that has been formed hydroplastically or by slip casting retains significant porosity and insufficient strength for most practical applications. In addition, it may still contain some liquid (e.g., water), which was added to assist in the forming operation. This liquid is removed in a drying process; density and strength are enhanced as a result of a high-temperature heat treatment or firing procedure. A body that has been formed and dried but not fired is termed **green**. Drying and firing techniques are critical inasmuch as defects that ordinarily render the ware useless (e.g., warpage, distortion, and cracks) may be introduced during the operation.

Summary

Crystal Structures

Both crystalline and noncrystalline states are possible for ceramics. The crystal structures of those materials for which the atomic bonding is predominantly ionic are determined by the charge magnitude and the radius of each kind of ion. Some of the simpler crystal structures are described in terms of unit cells; several of these were discussed (rock salt, cesium chloride, zinc blende, diamond cubic, graphite, fluorite, perovskite, and spinel structures).

Silicate Ceramics

For the silicates, structure is more conveniently represented by means of interconnecting SiO_4^{4-} tetrahedra. Relatively complex structures may result when other cations (e.g., Ca^{2+}, Mg^{2+}, Al^{3+}) and anions (e.g., OH^-) are added. The structures of silica (SiO_2), silica glass, and several of the simple and layered silicates were presented.

Carbon

The various forms of carbon-diamond, graphite, the fullerenes, and nanotubes-were also discussed. Diamond is a gem stone and, because of its hardness, is used to cut and grind softer materials. Furthermore, it is now being produced and utilized in thin films. The layered structure of graphite gives rise to its excellent lubricative properties and a relatively high electrical conductivity. Graphite is also known for its high strength and chemical stability at elevated temperatures and in nonoxidizing atmospheres. Fullerenes exist as hollow and spherical molecules composed of 60 carbon atoms. The recently discovered carbon nanotubes have extraordinary properties: high stiffnesses and strengths, low densities, and unusual electrical characteristics. Structurally, a nanotube is a graphite cylinder with fullerene hemispheres on its ends.

Brittle Fracture of Ceramics

Stress-Strain Behavior

At room temperature, virtually all ceramics are brittle. Microcracks, the presence of which is very difficult to control, result in amplification of applied tensile stresses and account for relatively low fracture strengths (flexural strengths). This amplification does not occur with compressive loads, and, consequently, ceramics are stronger in compression. Fractographic analysis of the fracture surface of a ceramic material may reveal the location and source of the crack-producing flaw, as well as the magnitude of the fracture stress. Representative strengths of ceramic materials are determined by performing transverse bending tests to fracture.

Glasses

Glass-Ceramics

This chapter began by discussing the various types of ceramic materials. The familiar glass materials are noncrystalline silicates that contain other oxides; the most desirable trait of these materials is their optical transparency. Glass-ceramics are initially fabricated as a glass, then crystallized.

Clay Products

Clay is the principal component of the whitewares and structural clay products. Other ingredients may be added, such as feldspar and quartz, which influence the changes that occur during firing.

Refractories

The materials that are employed at elevated temperatures and often in reactive environments are the refractory ceramics; on occasion, their ability to thermally insulate is also utilized. On the basis of composition and application, the four main subdivisions are fireclay, silica, basic, and special.

Abrasives

The abrasive ceramics, being hard and tough, are utilized to cut, grind, and polish other softer materials. Diamond, silicon carbide, tungsten carbide, corundum, and silica sand are the most common examples. The abrasives may be employed in the form of loose grains, bonded to an abrasive wheel, or coated on paper or a fabric.

Cements

When mixed with water, inorganic cements form a paste that is capable of assuming just about any desired shape. Subsequent setting or hardening is a result of chemical reactions involving the cement particles and occurs at the ambient temperature. For hydraulic cements, of which portland cement is the most common, the chemical reaction is one of hydration.

Advanced Ceramics

Many of our modern technologies utilize and will continue to utilize advanced ceramics because of their unique mechanical, chemical, electrical, magnetic, and optical properties and property combinations. The following advanced ceramic materials were discussed briefly: piezoelectric ceramics, microelectromechanical systems (MEMS), and ceramic ball bearings.

Fabrication and Processing of Ceramics

The next major section of this chapter discussed the principal techniques used for the fabrication of ceramic materials. Since glasses are formed at elevated temperatures, the temperature-viscosity behavior is an important consideration. Melting, working, softening, annealing, and strain points represent temperatures that correspond to specific viscosity values. Knowledge of these points is important in the fabrication and processing of a glass of given composition. Four of the more common glass-forming techniques—pressing, blowing, drawing, and fiber forming-were discussed briefly. After fabrication, glasses may be annealed and/or tempered to improve mechanical characteristics.

Fabrication and Processing of Clay Products

For clay products, two fabrication techniques that are frequently utilized are hydroplastic forming and slip casting. After forming, a body must be first dried and then fired at an elevated temperature to reduce porosity and enhance strength.

Shrinkage that is excessive or too rapid may result in cracking and/or warping, and a worthless piece. Densification during firing is accomplished by vitrification, the formation of a glassy bonding phase.

Important Terms and Concepts

Anion	Cation	Defect structure
Electroneutrality	Flexural strength	Frenkel defect
Octahedral position	Schottky defect	Stoichiometry
Tetrahedral position	Annealing point(glass)	Calcination
Abrasive(ceramic)	Crystallization(glass-ceramic)	Firing
Cement	Glass transition temperature	Green ceramic body
Glass-ceramic	Melting point(glass)	Refractory(ceramic)
Hydroplastic forming	Optical fiber	Softening point(glass)
Sintering	Slip casting	Thermal shock
Strain point(glass)	Structural clay product	Whiteware
Thermal tempering	Vitrification	Working point(glass)

References

[1] Barsoum, M. W., *Fundamentals of Ceramics*, McGraw-Hill, New York, 1997.
[2] Bergeron, C. G. and S. H. Risbud, *Introduction to Phase Equilibria in Ceramics*, American Ceramic Society, Columbus, OH, 1984.
[3] Bowen, H. K., "Advanced Ceramics," *Scientific American*, Vol. 255, No. 4, October 1986, pp. 168-176.
[4] Chiang, Y. M., D. P. Birnie, III, and W. D. Kingery, *Physical Ceramics: Principles for Ceramic Science and Engineering*, Wiley, New York, 1997.
[5] Curl, R. F. and R. E. Smalley, "Fullerenes," *Scientific American*, Vol. 265, No. 4, October 1991, pp. 54-63.
[6] Davidge, R. W., *Mechanical Behaviour of Ceramics*, Cambridge University Press, Cambridge, 1979. Reprinted by TechBooks, Marietta, OH, 1988.
[7] Doremus, R. H., *Glass Science*, 2nd edition, Wiley, New York, 1994.
[8] *Engineered Materials Handbook*, Vol. 4, *Ceramics and Glasses*, ASM International, Materials Park, OH, 1991.
[9] Green, D. J., *An Introduction to the Mechanical Properties of Ceramics*, Cambridge University Press, Cambridge, 1998.
[10] Kingery, W. D., H. K. Bowen, and D. R. Uhlmann, *Introduction to Ceramics*, 2nd edition, Wiley, New York, 1976. Chapters 1-4, 14, and 15.
[11] Norton, F. H., *Elements of Ceramics*, Addison-Wesley, 1974. Reprinted by TechBooks, Marietta, OH, 1991. Chapters 2 and 23.
[12] *Phase Equilibria Diagrams* (for Ceramists), American Ceramic Society, Westerville, OH. In fourteen volumes, published between 1964 and 2005.
[13] Richerson, D. W., *The Magic of Ceramics*, American Ceramic Society, Westerville, OH, 2000.
[14] Richerson, D. W., *Modern Ceramic Engineering*, 2nd edition, Marcel Dekker, New York, 1992.
[15] Wachtman, J. B., *Mechanical Properties of Ceramics*, Wiley, New York, 1996.
[16] *Engineered Materials Handbook*, Vol. 4, *Ceramics and Glasses*, ASM International, Materials Park, OH, 1991.
[17] Hewlett, P. C., *Lea's Chemistry of Cement & Concrete*, 4th edition, Butterworth-Heinemann, Woburn, UK, 2004.
[18] Kingery, W. D., H. K. Bowen, and D. R. Uhlmann, *Introduction to Ceramics*, 2nd edition, Wiley, New York, 1976. Chapters 1, 10, 11, and 16.
[19] Norton, F. H., *Refractories*, 4th edition. Reprinted by TechBooks, Marietta, OH, 1985.
[20] Reed, J. S., *Principles of Ceramic Processing*, 2nd edition, Wiley, New York, 1995.
[21] Richerson, D. W., *Modern Ceramic Engineering*, 2nd edition, Marcel Dekker, New York, 1992.
[22] Tooley, F. V. (Editor), *Handbook of Glass Manufacture*, Ashlee, New York, 1985. In two volumes.

Chapter 6   Composite Materials

Learning Objectives
After careful study of this chapter you should be able to do the following:
1. Name the three main divisions of composite materials, and cite the distinguishing feature of each.
2. Cite the difference in strengthening mechanism for large-particle and dispersion-strengthened particle-reinforced composites.
3. Note the three common fiber reinforcements used in polymer-matrix composites, and, for each, cite both desirable characteristics and limitations.
4. Cite the desirable features of metal-matrix composites.
5. Note the primary reason for the creation of ceramic-matrix composites.

6.1   Introduction

Many of our modern technologies require materials with unusual combinations of properties that cannot be met by the conventional metal alloys, ceramics, and polymeric materials. This is especially true for materials that are needed for aerospace, underwater, and transportation applications. For example, aircraft engineers are increasingly searching for structural materials that have low densities, are strong, stiff, and abrasion and impact resistant, and are not easily corroded. This is a rather formidable combination of characteristics. Frequently, strong materials are relatively dense; also, increasing the strength or stiffness generally results in a decrease in impact strength.

Material property combinations and ranges have been, and are yet being, extended by the development of composite materials. Generally speaking, a composite is considered to be any multiphase material that exhibits a significant proportion of the properties of both constituent phases such that a better combination of properties is realized. According to this **principle of combined action**, better property combinations are fashioned by the judicious combination of two or more distinct materials. Property trade-offs are also made for many composites.

Composites of sorts have already been discussed; these include multiphase metal alloys, ceramics, and polymers. For example, pearlitic steels have a microstructure consisting of alternating layers of α ferrite and cementite. The ferrite phase is soft and ductile, whereas cementite is hard and very brittle. The combined mechanical characteristics of the pearlite (reasonably high ductility and strength) are superior to those of either of the constituent phases. There are also a number of composites that occur in nature. For example, wood consists of strong and flexible cellulose fibers surrounded and held together by a stiffer material called lignin. Also, bone is a composite of the strong yet soft protein collagen and the hard, brittle mineral apatite.

A composite, in the present context, is a multiphase material that is artificially made, as opposed to one that occurs or forms naturally. In addition, the constituent phases must be chemically dissimilar and separated by a distinct interface. Thus, most metallic alloys and many ceramics do not fit this definition because their multiple phases are formed as a consequence of natural phenomena.

In designing composite materials, scientists and engineers have ingeniously combined various metals, ceramics, and polymers to produce a new generation of extraordinary materials. Most composites have been created to improve combinations of mechanical characteristics such as stiffness, toughness, and ambient and high-temperature strength.

Many composite materials are composed of just two phases; one is termed the **matrix**, which is continuous and surrounds the other phase, often called the **dispersed phase**. The properties of composites are a function of the properties of the constituent phases, their relative amounts, and the geometry of the dispersed phase. "Dispersed phase geometry" in this context means the shape of the particles and the particle size, distribution, and orientation.

One simple scheme for the classification of composite materials is shown in Figure 6.1, which consists of three main divisions—particle-reinforced, fiber-reinforced, and structural composites; also, at least two subdivisions exist for each. The dispersed phase for particle-reinforced composites is equiaxed (i.e., particle dimensions are approximately the same in all directions); for fiber-reinforced composites, the dispersed phase has the geometry of a fiber (i.e., a large length-to-diameter ratio). Structural composites are combinations of composites and homogenous materials. The discussion of the remainder of this chapter will be organized according to this classification scheme.

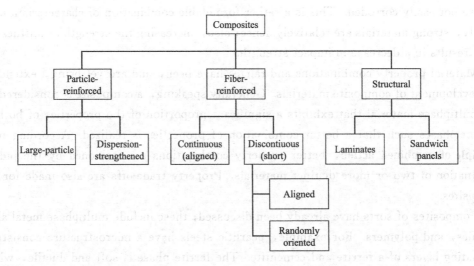

Figure 6.1 A classification scheme for the various composite types discussed in this chapter

6.2 Particle-Reinforced Composites

As noted in Figure 6.1, **large-particle** and **dispersion-strengthened composites** are the two subclassifications of particle-reinforced composites. The distinction between these is based upon reinforcement or strengthening mechanism. The term "large" is used to indicate that particle-matrix interactions cannot be treated on the atomic or molecular level; rather, continuum mechanics is used. For most of these composites, the particulate phase is harder and stiffer than the matrix. These reinforcing particles tend to restrain movement of the matrix phase in the vicinity of each particle. In essence, the matrix transfers some of the applied stress to the particles, which bear a fraction of the load. The degree of reinforcement or improvement of mechanical behavior depends on strong bonding at the matrix-particle interface. For dispersion-strengthened composites, particles are normally much smaller, having diameters between 0.01 and 0.1μm (10 and 100nm). Particle-matrix interactions that lead to strengthening occur on the atomic or molecular level. The mechanism of strengthening is similar to that for precipitation hardening discussed. Whereas the matrix bears the major portion of an applied load, the small dispersed particles hinder or impede the motion of dislocations. Thus, plastic deformation is restricted such that yield and tensile strengths, as well as hardness, improve.

6.2.1 Large-Particle Composites

Some polymeric materials to which fillers have been added are really large-particle composites. Again, the fillers modify or improve the properties of the material and/or replace some of the polymer volume with a less expensive material—the filler. Another familiar large-particle composite is concrete, being composed of cement (the matrix), and sand and gravel (the particulates).

Large-particle composites are utilized with all three material types (metals, polymers, and ceramics). The **cermets** are examples of ceramic—metal composites. The most common cermet is the cemented carbide, which is composed of extremely hard particles of a refractory carbide ceramic such as tungsten carbide (WC) or titanium carbide (TiC), embedded in a matrix of a metal such as cobalt or nickel. These composites are utilized extensively as cutting tools for hardened steels. The hard carbide particles provide the cutting surface but, being extremely brittle, are not themselves capable of withstanding the cutting stresses. Toughness is enhanced by their inclusion in the ductile metal matrix, which isolates the carbide particles from one another and prevents particle-to-particle crack propagation. Both matrix and particulate phases are quite refractory, to withstand the high temperatures generated by the cutting action on materials that are extremely hard. No single material could possibly provide the combination of properties possessed by a cermet. Relatively large volume fractions of the particulate phase may be utilized, often exceeding 90 vol%; thus the abrasive action of the composite is maximized.

Both elastomers and plastics are frequently reinforced with various particulate materials. Our use of many of the modern rubbers would be severely restricted without reinforcing particulate materials such as carbon black. Carbon black consists of very small and essentially spherical particles of carbon, produced by the combustion of natural gas or oil in an atmosphere that has only a limited air supply. When added to vulcanized rubber, this extremely inexpensive material enhances tensile strength, toughness, and tear and abrasion resistance. Automobile tires contain on the order of 15 to 30 vol% of carbon black. For the carbon black to provide significant reinforcement, the particle size must be extremely small, with diameters between 20 and 50 nm; also, the particles must be evenly distributed throughout the rubber and must form a strong adhesive bond with the rubber matrix. Particle reinforcement using other materials (e.g., silica) is much less effective because this special interaction between the rubber molecules and particle surfaces does not exist.

6.2.2 Dispersion-Strengthened Composites

Metals and metal alloys may be strengthened and hardened by the uniform dispersion of several volume percent of fine particles of a very hard and inert material. The dispersed phase may be metallic or nonmetallic; oxide materials are often used. Again, the strengthening mechanism involves interactions between the particles and dislocations within the matrix, as with precipitation hardening. The dispersion strengthening effect is not as pronounced as with precipitation hardening; however, the strengthening is retained at elevated temperatures and for extended time periods because the dispersed particles are chosen to be unreactive with the matrix phase. For precipitation-hardened alloys, the increase in strength may disappear upon heat treatment as a consequence of precipitate growth or dissolution of the precipitate phase.

The high-temperature strength of nickel alloys may be enhanced significantly by the addition of about 3 vol% of thoria (ThO_2) as finely dispersed particles; this material is known as thoria-dispersed (or TD) nickel. The same effect is produced in the aluminum—aluminum oxide system. A very thin and adherent alumina coating is caused to form on the surface of extremely small (0.1 to 0.2 μm thick) flakes of aluminum, which are dispersed within an aluminum metal matrix; this material is termed sintered aluminum powder (SAP).

6.3 Fiber-Reinforced Composites

Technologically, the most important composites are those in which the dispersed phase is in the form of a fiber. Design goals of **fiber-reinforced composites** often include high strength and/or stiffness on a weight basis. These characteristics are expressed in terms of **specific strength** and **specific modulus** parameters, which correspond, respectively, to the ratios of tensile strength to specific gravity and modulus of elasticity to specific gravity. Fiber-reinforced composites with exceptionally high specific strengths and moduli have been produced that utilize low-density fiber and matrix materials.

As noted in Figure 6.1, fiber-reinforced composites are subclassified by fiber length. For short fiber, the fibers are too short to produce a significant improvement in strength.

6.3.1 The Fiber Phase

An important characteristic of most materials, especially brittle ones, is that a small diameter fiber is much stronger than the bulk material. The probability of the presence of a critical surface flaw that can lead to fracture diminishes with decreasing specimen volume, and this feature is used to advantage in the fiber-reinforced composites. Also, the materials used for reinforcing fibers have high tensile strengths.

On the basis of diameter and character, fibers are grouped into three different classifications: *whiskers*, *fibers*, and *wires*. **Whiskers** are very thin single crystals that have extremely large length-to-diameter ratios. As a consequence of their small size, they have a high degree of crystalline perfection and are virtually flaw free, which accounts for their exceptionally high strengths; they are the strongest known materials. In spite of these high strengths, whiskers are not utilized extensively as a reinforcement medium because they are extremely expensive. Moreover, it is difficult and often impractical to incorporate whiskers into a matrix. Whisker materials include graphite, silicon carbide, silicon nitride, and aluminum oxide.

Materials that are classified as **fibers** are either polycrystalline or amorphous and have small diameters; fibrous materials are generally either polymers or ceramics (e.g., the polymer aramids, glass, carbon, boron, aluminum oxide, and silicon carbide). Fine wires have relatively large diameters; typical materials include steel, molybdenum, and tungsten. Wires are utilized as a radial steel reinforcement in automobile tires, in filament-wound rocket casings, and in wire-wound high-pressure hoses.

6.3.2 The Matrix Phase

The **matrix phase** of fibrous composites may be a metal, polymer, or ceramic. In general, metals and polymers are used as matrix materials because some ductility is desirable; for ceramic-matrix composites, the reinforcing component is added to improve fracture toughness. The discussion of this section will focus on polymer and metal matrices.

For fiber-reinforced composites, the matrix phase serves several functions. First, it binds the fibers together and acts as the medium by which an externally applied stress is transmitted and distributed to the fibers; only a very small proportion of an applied load is sustained by the matrix phase. Furthermore, the matrix material should be ductile. In addition, the elastic modulus of the fiber should be much higher than that of the matrix. The second function of the matrix is to protect the individual fibers from surface damage as a result of mechanical abrasion or chemical reactions with the environment. Such interactions may introduce surface flaws capable of forming cracks, which may lead to failure at low tensile stress levels. Finally, the matrix separates the fibers and, by virtue of its relative softness and plasticity, prevents the propagation of brittle cracks from fiber to fiber, which could result in catastrophic failure; in other words, the matrix phase serves as a barrier to crack propagation. Even though some of the individual fibers fail, total composite fracture will not occur until large numbers of adjacent fibers, once having failed, form a cluster of critical size.

It is essential that adhesive bonding forces between fiber and matrix be high to minimize fiber pull-out. In fact, bonding strength is an important consideration in the choice of the matrix—fiber combination. The ultimate strength of the composite depends to a large degree on the magnitude of this bond; adequate bonding is essential to maximize the stress transmittance from the weak matrix to the strong fibers.

6.4 Polymer-Matrix Composites

Polymer-matrix composites (PMCs) consist of a polymer resin1 as the matrix, with fibers as the reinforcement medium. These materials are used in the greatest diversity of composite applications, as well as in the largest quantities, in light of their room-temperature properties, ease of fabrication, and cost. In this section the various classifications of PMCs are discussed according to reinforcement type (i.e., glass, carbon, and aramid), along with their applications and the various polymer resins that are employed.

6.4.1 Glass Fiber-Reinforced Polymer (GFRP) Composites

Fiberglass is simply a composite consisting of glass fibers, either continuous or discontinuous, contained within a polymer matrix; this type of composite is produced in the largest quantities.

Glass is popular as a fiber reinforcement material for several reasons:
1. It is easily drawn into high-strength fibers from the molten state.
2. It is readily available and may be fabricated into a glass-reinforced plastic economically using a wide variety of composite-manufacturing techniques.
3. As a fiber, it is relatively strong, and when embedded in a plastic matrix, it produces a composite having a very high specific strength.
4. When coupled with the various plastics, it possesses a chemical inertness that renders the composite useful in a variety of corrosive environments.

The surface characteristics of glass fibers are extremely important because even minute surface flaws can deleteriously affect the tensile properties. Surface flaws are easily introduced by rubbing or abrading the surface with another hard material. Also, glass surfaces that have been exposed to the normal atmosphere for even short time periods generally have a weakened surface layer that interferes with bonding to the matrix. Newly drawn fibers are normally coated during drawing with a "size", a thin layer of a substance that protects the fiber surface from damage and undesirable environmental interactions. This size is ordinarily removed prior to composite fabrication and replaced with a "coupling agent" or finish that promotes a better bond between the fiber and matrix.

There are several limitations to this group of materials. In spite of having high strengths, they are not very stiff and do not display the rigidity that is necessary for some applications (e.g., as structural members for airplanes and bridges). Most fiberglass materials are limited to service temperatures below 200℃ (400°F); at higher temperatures, most polymers begin to flow or to deteriorate. Service temperatures may be extended to approximately 300℃ (575°F) by using high-purity fused silica for the fibers and high-temperature polymers such as the polyimide resins.

Many fiberglass applications are familiar: automotive and marine bodies, plastic pipes, storage containers, and industrial floorings. The transportation industries are utilizing increasing amounts of glass fiber-reinforced plastics in an effort to decrease vehicle weight and boost fuel efficiencies. A host of new applications are being used or currently investigated by the automotive industry.

6.4.2 Carbon Fiber-Reinforced Polymer (CFRP) Composites

Carbon is a high-performance fiber material that is the most commonly used reinforcement in advanced (i.e., nonfiberglass) polymer-matrix composites. The reasons for this are as follows:

1. Carbon fibers have the highest specific modulus and specific strength of all reinforcing fiber materials.
2. They retain their high-tensile modulus and high strength at elevated temperatures; high-temperature oxidation, however, may be a problem.
3. At room temperature carbon fibers are not affected by moisture or a wide variety of solvents, acids, and bases.
4. These fibers exhibit a diversity of physical and mechanical characteristics, allowing composites incorporating these fibers to have specific engineered properties.
5. Fiber and composite manufacturing processes have been developed that are relatively inexpensive and cost effective.

Use of the term "carbon fiber" may seem perplexing inasmuch as carbon is an element, and the stable form of crystalline carbon at ambient conditions is graphite. Carbon fibers are not totally crystalline, but are composed of both graphitic and noncrystalline regions; these areas of noncrystallinity are devoid of the three-dimensional ordered arrangement of hexagonal carbon networks that is characteristic of graphite.

Manufacturing techniques for producing carbon fibers are relatively complex and will not be discussed. However, three different organic precursor materials are used—rayon, polyacrylonitrile (PAN), and pitch. Processing technique will vary from precursor to precursor, as will also the resultant fiber characteristics.

One classification scheme for carbon fibers is according to tensile modulus; on this basis the four classes are standard, intermediate, high, and ultrahigh moduli. Furthermore, fiber diameters normally range between 4 and 10 μm; both continuous and chopped forms are available. In addition, carbon fibers are normally coated with a protective epoxy size that also improves adhesion with the polymer matrix.

Carbon-reinforced polymer composites are currently being utilized extensively in sports and recreational equipment (fishing rods, golf clubs), filament-wound rocket motor cases, pressure vessels, and aircraft structural components—both military and commercial, fixed wing and helicopters (e.g., as wing, body, stabilizer, and rudder components).

6.4.3 Aramid Fiber-Reinforced Polymer Composites

Aramid fibers are high-strength, high-modulus materials that were introduced in the early 1970s. They are especially desirable for their outstanding strength-toweight ratios, which are superior to metals. Chemically, this group of materials is known as poly paraphenylene terephthalamide. There are a number of aramid materials; trade names for two of the most common are Kevlar and Nomex. For the former, there are several grades (viz. Kevlar 29, 49, and 149) which have different mechanical behaviors. During synthesis, the rigid molecules are aligned in the direction of the fiber axis, as liquid crystal domains; the mer chemistry and mode of chain alignment are represented in Figure 6.2. Mechanically, these fibers have longitudinal tensile strengths and tensile moduli that are higher than other polymeric fiber materials; however, they are relatively weak in compression. In addition, this material is known for its toughness, impact resistance, and resistance to creep and fatigue failure. Even though the aramids are thermoplastics, they are, nevertheless, resistant to combustion and stable to relatively high temperatures; the temperature range over which they retain their high mechanical properties is between $-200\,℃$ and $200\,℃$ ($-330\,℉$ and $390\,℉$). Chemically, they are susceptible to degradation by strong acids and bases, but they are relatively inert in other solvents and chemicals.

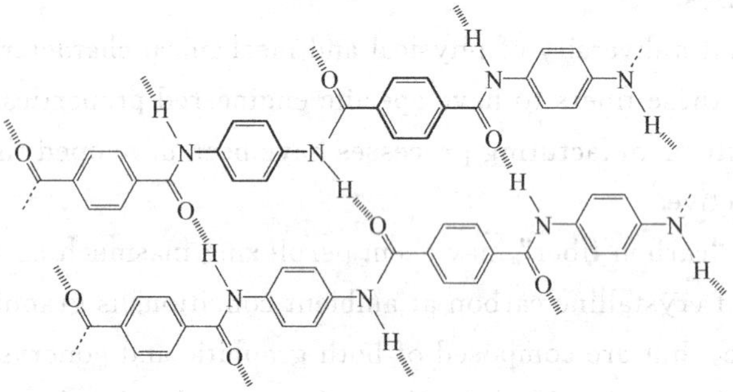

Figure 6.2 Schematic representation of repeat unit and chain structures for aramid (Kevlar) fibers. Chain alignment with the fiber direction and hydrogen bonds that form between adjacent chains are also shown

The aramid fibers are most often used in composites having polymer matrices; common matrix materials are the epoxies and polyesters. Since the fibers are relatively flexible and somewhat ductile, they may be processed by most common textile operations. Typical applications of these aramid composites are in ballistic products (bullet-proof vests), sporting goods, tires, ropes, missile cases, pressure vessels, and as a replacement for asbestos in automotive brake and clutch linings, and gaskets.

The properties of continuous and aligned glass-, carbon-, and aramid-fiber reinforced epoxy composites are included in Table 6.1. Thus, a comparison of the mechanical characteristics of these three materials may be made in both longitudinal and transverse directions.

In addition, the matrix often determines the maximum service temperature, since it normally softens, melts, or degrades at a much lower temperature than the fiber reinforcement. The most widely utilized and least expensive polymer resins are the polyesters and vinyl esters; these matrix materials are used primarily for glass fiber-reinforced composites. A large number of resin formulations provide a wide range of properties for these polymers. The epoxies are more expensive and, in addition to commercial applications, are also utilized extensively in PMCs for aerospace applications; they have better mechanical properties and resistance to moisture than the polyesters and vinyl resins. For high-temperature applications, polyimide resins are employed; their continuous-use, upper-temperature limit is approximately 230℃ (450°F). And finally, high-temperature thermoplastic resins offer the potential to be used in future aerospace applications; such materials include polyetheretherketone (PEEK), polyphenylene sulfide (PPS), and polyetherimide (PEI).

Table 6.1 Properties of Continuous and Aligned Glass-, Carbon-, and Aramid-Fiber Reinforced Epoxy-Matrix Composites in Longitudinal and Transverse Directions. In All Cases the Fiber Volume Fraction Is 0.60

Property	Glass(E-glass)	Carbon(High Strength)	Aramid(Kevlar 49)
Specific gravity	2.1	1.6	1.4
Tensile modulus	—	—	—
Longitudinal [GPa(10^6 psi)]	45(6.5)	145(21)	76(11)
Transverse [GPa(10^6 psi)]	12(1.8)	10(1.5)	5.5(0.8)
Tensile strength	—	—	—
Longitudinal [MPa(ksi)]	1020(150)	1240(180)	1380(200)
Transverse [MPa ksi)]	40(5.8)	41(6)	30(4.3)
Ultimate tensile strain	—	—	—
Longitudinal	2.3	0.9	1.8
Transverse	0.4	0.4	0.5

6.5 Metal-Matrix Composites

As the name implies, for **metal-matrix composites** (MMCs), the matrix is a ductile metal. These materials may be utilized at higher service temperatures than their base metal counterparts; furthermore, the reinforcement may improve specific stiffness, specific strength, abrasion resistance, creep resistance, thermal conductivity, and dimensional stability. Some of the advantages of these materials over the polymer matrix composites include higher operating temperatures, nonflammability, and greater resistance to degradation by organic fluids. Metal-matrix composites are much more expensive than PMCs, and, therefore, their (MMC) use is somewhat restricted.

The superalloys, as well as alloys of aluminum, magnesium, titanium, and copper, are employed as matrix materials. The reinforcement may be in the form of particulates, both continuous and discontinuous fibers, and whiskers; concentrations normally range between 10 and 60 vol%. Continuous fiber materials include carbon, silicon carbide, boron, alumina, and the refractory metals. On the other hand, discontinuous reinforcements consist primarily of silicon carbide whiskers, chopped fibers of alumina and carbon, and particulates of silicon carbide and alumina. In a sense, the cermets fall within this MMC scheme. In Table 6.2 are presented the properties of several common metal-matrix, continuous and aligned fiber-reinforced composites.

Some matrix-reinforcement combinations are highly reactive at elevated temperatures. Consequently, composite degradation may be caused by high-temperature processing, or by subjecting the MMC to elevated temperatures during service. This problem is commonly resolved either by applying a protective surface coating to the reinforcement or by modifying the matrix alloy composition. Normally the processing of MMCs involves at least two steps: consolidation or synthesis (i. e., introduction of reinforcement into the matrix), followed by a shaping operation. A host of consolidation techniques are available, some of which are relatively sophisticated; discontinuous fiber MMCs are amenable to shaping by standard metal-forming operations (e. g., forging, extrusion, rolling).

Table 6.2 Properties of Several Metal-Matrix Composites Reinforced with Continuous and Aligned Fibers

Fiber	Matrix	Fiber Content /(vol%)	Density /(g/cm^3)	Longitudinal Tensile Modulus/GPa	Longitudinal Tensile Strength/MPa
Cabon	6061 Al	41	2.44	320	620
Boron	6061 Al	48	—	207	1515
SiC	6061 Al	50	2.93	230	1480
Alumina	380.0 Al	24	—	120	340
Carbon	AZ31 Mg	38	1.83	300	510
Borsic	Ti	45	3.68	220	1270

Recently, some of the automobile manufacturers have introduced engine components consisting of an aluminum-alloy matrix that is reinforced with alumina and carbon fibers; this MMC is light in weight and resists wear and thermal distortion. Aerospace structural applications include advanced aluminum alloy metal-matrix composites; boron fibers are used as the reinforcement for the Space Shuttle Orbiter, and continuous graphite fibers for the Hubble Telescope.

The high-temperature creep and rupture properties of some of the superalloys (Ni- and Co-based alloys) may be enhanced by fiber reinforcement using refractory metals such as tungsten. Excellent high-temperature oxidation resistance and impact strength are also maintained. Designs incorporating these composites permit higher operating temperatures and better efficiencies for turbine engines.

6.6 Ceramic-Matrix Composites

Ceramic materials are inherently resilient to oxidation and deterioration at elevated temperatures; were it not for their disposition to brittle fracture, some of these materials would be ideal candidates for use in high temperature and severe-stress applications, specifically for components in automobile and aircraft gas turbine engines. Fracture toughness values for ceramic materials are low and typically lie between 1 and 5 MPa $\sqrt{m}$ (0.9 and 4.5 ksi $\sqrt{in.}$). By way of contrast, K_{Ic} values for most metals are much higher (15 to greater than 150 MPa $\sqrt{m}$ [14 to >140 ksi $\sqrt{in.}$]).

The fracture toughnesses of ceramics have been improved significantly by the development of a new generation of **ceramic-matrix composites** (CMCs)—particulates, fibers, or whiskers of one ceramic material that have been embedded into a matrix of another ceramic. Ceramic-matrix composite materials have extended fracture toughnesses to between about 6 and 20MPa $\sqrt{m}$ (5.5 and 18 ksi $\sqrt{in.}$). In essence, this improvement in the fracture properties results from interactions between advancing cracks and dispersed phase particles. Crack initiation normally occurs with the matrix phase, whereas crack propagation is impeded or hindered by the particles, fibers, or whiskers. Several techniques are utilized to retard crack propagation, which are discussed as follows.

One particularly interesting and promising toughening technique employs a phase transformation to arrest the propagation of cracks and is aptly termed *transformation toughening*. Small particles of partially stabilized zirconia are dispersed within the matrix material, often Al_2O_3 or ZrO_2 itself. Typically, CaO, MgO, Y_2O_3, and CeO are used as stabilizers. Partial stabilization allows retention of the metastable tetragonal phase at ambient conditions rather than the stable monoclinic phase; these two phases are noted on the ZrO_2-$ZrCaO_3$ phase diagram. The stress field in front of a propagating crack causes these metastably retained tetragonal particles to undergo transformation to the stable monoclinic phase. Accompanying this transformation is a slight particle volume increase, and the net result is that compressive stresses are established on the crack surfaces near the crack tip that tend to pinch the crack shut, thereby arresting its growth. This process is demonstrated schematically in Figure 6.3.

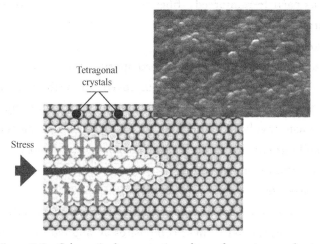

Figure 6.3 Schematic demonstration of transformation toughening

Other recently developed toughening techniques involve the utilization of ceramic whiskers, often SiC or Si_3N_4. These whiskers may inhibit crack propagation by (1) deflecting crack tips, (2) forming bridges across crack faces, (3) absorbing energy during pull-out as the whiskers debond from the matrix, and/or (4) causing a redistribution of stresses in regions adjacent to the crack tips.

In general, increasing fiber content improves strength and fracture toughness; this is demonstrated in Table 6.3 for SiC whisker-reinforced alumina. Furthermore, there is a considerable reduction in the scatter of fracture strengths for whiskerreinforced ceramics relative to their unreinforced counterparts. In addition, these CMCs exhibit improved high-temperature creep behavior and resistance to thermal shock (i.e., failure resulting from sudden changes in temperature).

Ceramic-matrix composites may be fabricated using hot pressing, hot isostatic pressing, and liquid phase sintering techniques. Relative to applications, SiC whisker-reinforced aluminas are being utilized as cutting tool inserts for machining hard metal alloys; tool lives for these materials are greater than for cemented carbides.

Table 6.3　Room-Temperature Fracture Strengths and
Fracture Toughnesses for Various SiC Whisker Contents in Al_2O_3

Whisker Content/(vol%)	Fracture Strength/MPa	Fracture Toughness/MPa $\sqrt{m}$
0	—	4.5
10	455 ± 55	7.1
20	655 ± 135	7.5~9.0
40	850 ± 130	6.0

Summary
Introduction

Composites are artificially produced multiphase materials having a desirable combination of the best properties of the constituent phases. Usually, one phase (the matrix) is continuous and completely surrounds the other (the dispersed phase). In this discussion, composites were classified as particle-reinforced, fiber-reinforced, and structural composites.

Large-Particle Composites
Dispersion-Strengthened Composites

Large-particle and dispersion-strengthened composites fall within the particle reinforced classification. For dispersion strengthening, improved strength is achieved by extremely small particles of the dispersed phase, which inhibit dislocation motion; that is, the strengthening mechanism involves interactions that may be treated on the atomic level. The particle size is normally greater with large particle composites, whose mechanical characteristics are enhanced by reinforcement action.

Fiber-Reinforced Composites

Of the several composite types, the potential for reinforcement efficiency is greatest for those that are fiber reinforced. With these composites an applied load is transmitted and distributed among the fibers via the matrix phase, which in most cases is at least moderately ductile. Significant reinforcement is possible only if the matrix-fiber bond is strong. On the basis of diameter, fiber reinforcements are classified as whiskers, fibers, or wires. Since reinforcement discontinues at the fiber extremities, reinforcement efficiency depends on fiber length. For each fiber—matrix combination, there exists some critical length; the length of continuous fibers greatly exceeds this critical value, whereas shorter fibers are discontinuous.

Fiber arrangement is also crucial relative to composite characteristics. The mechanical properties of continuous and aligned fiber composites are highly anisotropic. In the alignment direction, reinforcement and strength are a maximum; perpendicular to the alignment, they are a minimum. The stress-strain behavior for longitudinal loading was discussed. Composite rule-of-mixture expressions for the modulus in both longitudinal and transverse orientations were developed; in addition, an equation for longitudinal strength was also cited.

For short and discontinuous fibrous composites, the fibers may be either aligned or randomly oriented. Significant strengths and stiffnesses are possible for aligned short-fiber composites in the longitudinal direction. Despite some limitations on reinforcement efficiency, the properties of randomly oriented short-fiber composites are isotropic.

Polymer-Matrix Composites
Metal-Matrix Composites
Ceramic-Matrix Composites

Fibrous-reinforced composites are sometimes classified according to matrix type; within this scheme are three classifications—viz. polymer-, metal-, and ceramic-matrix. Polymer-matrix is the most common, which may be reinforced with glass, carbon, and aramid fibers. Service temperatures are higher for metal-matrix composites, which also utilize a variety of fiber and whisker types. The objective of many polymer- and metal-matrix composites is a high specific strength and/or specific modulus, which requires matrix materials having low densities. With ceramic-matrix composites, the design goal is increased fracture toughness. This is achieved by interactions between advancing cracks and dispersed phase particles; transformation toughening is one such technique for improving K_{Ic}. Other more advanced composites are carbon-carbon (carbon fibers embedded in a pyrolyzed carbon matrix) and the hybrids (containing at least two different fiber types).

References

[1] Agarwal, B. D. and L. J. Broutman, *Analysis and Performance of Fiber Composites*, 2nd edition, Wiley, New York, 1990.
[2] Ashbee, K. H., *Fundamental Principles of Fiber Reinforced Composites*, 2nd edition, Technomic Publishing Company, Lancaster, PA, 1993.
[3] *ASM Handbook*, Vol. 21, *Composites*, ASM International, Materials Park, OH, 2001.
[4] Chawla, K. K., *Composite Materials Science and Engineering*, 2nd edition, Springer-Verlag, New York, 1998.
[5] Chou, T. W., R. L. McCullough, and R. B. Pipes, "Composites," *Scientific American*, Vol. 255, No. 4, October 1986, pp. 192-203.
[6] Hollaway, L. (Editor), *Handbook of Polymer Composites for Engineers*, Technomic Publishing Company, Lancaster, PA, 1994.
[7] Hull, D. and T. W. Clyne, *An Introduction to Composite Materials*, 2nd edition, Cambridge University Press, New York, 1996.
[8] Mallick, P. K., *Fiber-Reinforced Composites, Materials, Manufacturing, and Design*, 2nd edition, Marcel Dekker, New York, 1993.
[9] Peters, S. T., *Handbook of Composites*, 2nd edition, Springer-Verlag, New York, 1998.
[10] Strong, A. B., *Fundamentals of Composites: Materials, Methods, and Applications*, Society of Manufacturing Engineers, Dearborn, MI, 1989.
[11] Woishnis, W. A. (Editor), *Engineering Plastics and Composites*, 2nd edition, ASM International, Materials Park, OH, 1993.
[12] William D. Callister Jr., *Fundamentals of Materials Science and Engineering*, 7th Edition, John Wiley & Sons, Inc., 2007.

Chapter 7 Corrosion and Degradation of Materials

Learning Objectives

After careful study of this chapter you should be able to do the following:
1. Distinguish between oxidation and reduction electrochemical reactions.
2. Determine metal corrosion rate given the reaction current density.
3. List five measures that are commonly used to prevent corrosion.
4. Explain why ceramic materials are, in general, very resistant to corrosion.
5. For polymeric materials discuss (a) two degradation processes that occur when they are exposed to liquid solvents, and (b) the causes and consequences of molecular chain bond rupture.

7.1 Introduction

To one degree or another, most materials experience some type of interaction with a large number of diverse environments. Often, such interactions impair a material's usefulness as a result of the deterioration of its mechanical properties (e.g., ductility and strength), other physical properties, or appearance.

Deteriorative mechanisms are different for the three material types. In metals, there is actual material loss either by dissolution (corrosion) or by the formation of nonmetallic scale or film (oxidation). Ceramic materials are relatively resistant to deterioration, which usually occurs at elevated temperatures or in rather extreme environments; the process is frequently also called corrosion. For polymers, mechanisms and consequences differ from those for metals and ceramics, and the term degradation is most frequently used. Polymers may dissolve when exposed to a liquid solvent, or they may absorb the solvent and swell; also, electromagnetic radiation (primarily ultraviolet) and heat may cause alterations in their molecular structures.

7.2 Corrosion of Metals

Corrosion is defined as the destructive and unintentional attack of a metal; it is electrochemical and ordinarily begins at the surface. The problem of metallic corrosion is one of significant proportions; in economic terms, it has been estimated that approximately 5% of an industrialized nation's income is spent on corrosion prevention and the maintenance or replacement of products lost or contaminated as a result of corrosion reactions. The consequences of corrosion are all too common. Familiar examples include the rusting of automotive body panels and radiator and exhaust components.

7.2.1 Electrochemical Considerations

For metallic materials, the corrosion process is normally electrochemical, that is, a chemical reaction in which there is transfer of electrons from one chemical species to another. Metal atoms characteristically lose or give up electrons in what is called an oxidation reaction. Examples in which metals oxidize are

$$Fe \longrightarrow Fe^{2+} + 2e^-, \qquad (7.1a)$$

$$Al \longrightarrow Al^{3+} + 3e^-. \qquad (7.1b)$$

The site at which oxidation takes place is called the **anode**. The electrons generated from each metal atom that is oxidized must be transferred to and become a part of another chemical species in what is termed a **reduction reaction**. For example, some metals undergo corrosion in acid solutions, which have a high concentration of hydrogen (H^+) ions; the H^+ ions are reduced as hydrogen gas. Other reduction reactions are possible, depending on the nature of the solution to which the metal is exposed. For an acid solution having dissolved oxygen, reduction according to

$$O_2 + 4H^+ + 4e^- \longrightarrow 2H_2O, \qquad (7.2)$$

will probably occur. Or, for a neutral or basic aqueous solution in which oxygen is also dissolved,

$$O_2 + 2H_2O + 4e^- \longrightarrow 4(OH^-). \qquad (7.3)$$

The location at which reduction occurs is called the **cathode**. Furthermore, it is possible for two or more of the reduction reactions above to occur simultaneously.

As a consequence of oxidation, the metal ions may either go into the corroding solution as ions, or they may form an insoluble compound with nonmetallic elements.

Not all metallic materials oxidize to form ions with the same degree of ease. Metallic materials may be rated as to their tendency to experience oxidation when coupled to other metals in solutions of their respective ions. Table 7.1 represents the corrosion tendencies for the several metals; those at the top (i.e., gold and platinum) are noble, or chemically inert. Moving down the table, the metals become increasingly more active, that is, more susceptible to oxidation. Sodium and potassium have the highest reactivities.

Most metals and alloys are subject to oxidation or corrosion to one degree or another in a wide variety of environments; that is, they are more stable in an ionic state than as metals. In thermodynamic terms, there is a net decrease in free energy in going from metallic to oxidized states. Consequently, essentially all metals occur in nature as compounds—for example, oxides, hydroxides, carbonates, silicates, sulfides, and sulfates. Two notable exceptions are the noble metals gold and platinum. For them, oxidation in most environments is not favorable, and, therefore, they may exist in nature in the metallic state.

Even though Table 7.1 was generated under highly idealized conditions and has limited utility, it nevertheless indicates the relative reactivities of the metals. A more realistic and practical ranking, however, is provided by the galvanic series, Table 7.2. This represents the relative reactivities of a number of metals and commercial alloys in seawater. The alloys near the top are cathodic and unreactive, whereas those at the bottom are most anodic; no voltages are provided.

Table 7.1 The Corrosion Tendencies for Common Metals

	Electrode Reaction
↑	$Au^{3+} + 3e^- \longrightarrow Au$
	$O_2 + 4H^+ + 4e^- \longrightarrow 2H_2O$
	$Pt^{2+} + 2e^- \longrightarrow Pt$
	$Ag^+ + e^- \longrightarrow Ag$
	$Fe^{3+} + e^- \longrightarrow Fe^{2+}$
	$O_2 + 2H_2O + 4e^- \longrightarrow 4(OH^-)$
	$Cu^{2+} + 2e^- \longrightarrow Cu$
	$2H^+ + 2e^- \longrightarrow H_2$
Increasingly inert (cathodic)	$Pb^{2+} + 2e^- \longrightarrow Pb$
	$Sn^{2+} + 2e^- \longrightarrow Sn$
	$Ni^{2+} + 2e^- \longrightarrow Ni$
Increasingly active (anodic)	$Co^{2+} + 2e^- \longrightarrow Co$
	$Cd^{2+} + 2e^- \longrightarrow Cd$
	$Fe^{2+} + 2e^- \longrightarrow Fe$
	$Cr^{3+} + 3e^- \longrightarrow Cr$
	$Zn^{2+} + 2e^- \longrightarrow Zn$
	$Al^{3+} + 2e^- \longrightarrow Al$
	$Mg^{2+} + 2e^- \longrightarrow Mg$
	$Na^+ + e^- \longrightarrow Na$
↓	$K^+ + e^- \longrightarrow K$

Table 7.2 The Galvanic Series

↑	Platinum
	Gold
	Graphite
	Titanium
	Silver
	⌈ 316 Stainless steel (passive)
	⌊ 304 Stainless steel (passive)
	⌈ Inconel (80Ni-13Cr-7Fe) (passive)
	⌊ Nickel (passive)
	⌈ Monel (70Ni-30Cu)
	Copper-nickel alloys
Increasingly inert (cathodic)	Bronzes (Cu-Sn alloys)
	Copper
	⌊ Brasses (Cu-Zn alloys)
	⌈ Inconel (active)
Increasingly active (anodic)	⌊ Nickel (active)
	Tin
	Lead
	⌈ 316 Stainless steel (active)
	⌊ 304 Stainless steel (active)
	⌈ Cast iron
	⌊ Iron and steel
	Aluminum alloys
	Cadmium
	Commercially pure aluminum
	Zinc
↓	Magnesium and magnesium alloys

· 152 ·

7.2.2 Corrosion Rates

The corrosion rate, or the rate of material removal as a consequence of the chemical action, is an important corrosion parameter. This may be expressed as the **corrosion penetration rate** (CPR), or the thickness loss of material per unit of time. The formula for this calculation is

$$\text{CPR} = \frac{KW}{\rho A t} \tag{7.4}$$

where W is the weight loss after exposure time t; ρ and A represent the density and exposed specimen area, respectively, and K is a constant, its magnitude depending on the system of units used. The CPR is conveniently expressed in terms of millimeters per year (mm/a).

7.2.3 Passivity

Some normally active metals and alloys, under particular environmental conditions, lose their chemical reactivity and become extremely inert. This phenomenon, termed **passivity**, is displayed by chromium, iron, nickel, titanium, and many of their alloys. It is felt that this passive behavior results from the formation of a highly adherent and very thin oxide film on the metal surface, which serves as a protective barrier to further corrosion. Stainless steels are highly resistant to corrosion in a rather wide variety of atmospheres as a result of passivation. They contain at least 11% chromium that, as a solid-solution alloying element in iron, minimizes the formation of rust; instead, a protective surface film forms in oxidizing atmospheres. (Stainless steels are susceptible to corrosion in some environments, and therefore are not always "stainless".) Aluminum is highly corrosion resistant in many environments because it also passivates. If damaged, the protective film normally reforms very rapidly. However, a change in the character of the environment (e.g., alteration in the concentration of the active corrosive species) may cause a passivated material to revert to an active state. Subsequent damage to a preexisting passive film could result in a substantial increase in corrosion rate, by as much as 100000 times.

7.2.4 Environmental Effects

The variables in the corrosion environment, which include fluid velocity, temperature, and composition, can have a decided influence on the corrosion properties of the materials that are in contact with it. In most instances, increasing fluid velocity enhances the rate of corrosion due to erosive effects. The rates of most chemical reactions rise with increasing temperature; this also holds for the great majority of corrosion situations. Increasing the concentration of the corrosive species (e.g., H^+ ions in acids) in many situations produces a more rapid rate of corrosion. However, for materials capable of passivation, raising the corrosive content may result in an active-to-passive transition, with a considerable reduction in corrosion.

Cold working or plastically deforming ductile metals is used to increase their strength; however, a cold-worked metal is more susceptible to corrosion than the same material in an annealed state. For example, deformation processes are used to shape the head and point of a nail; consequently, these positions are anodic with respect to the shank region.

7.2.5 Forms of Corrosion

It is convenient to classify corrosion according to the manner in which it is manifest. Metallic corrosion is sometimes classified into eight forms: uniform, galvanic, crevice, pitting, intergranular, selective leaching, erosion-corrosion, and stress corrosion. In addition, hydrogen embrittlement is, in a strict sense, a type of failure rather than a form of corrosion; however, it is often produced by hydrogen that is generated from corrosion reactions.

Uniform attack

Uniform attack is a form of electrochemical corrosion that occurs with equivalent intensity over the entire exposed surface and often leaves behind a scale or deposit. In a microscopic sense, the oxidation and reduction reactions occur randomly over the surface. Some familiar examples include general rusting of steel and iron and the tarnishing of silverware. This is probably the most common form of corrosion. It is also the least objectionable because it can be predicted and designed for with relative ease.

Galvanic corrosion

Galvanic corrosion occurs when two metals or alloys having different compositions are electrically coupled while exposed to an electrolyte. The less noble or more reactive metal in the particular environment will experience corrosion; the more inert metal, the cathode, will be protected from corrosion. For example, steel screws corrode when in contact with brass in a marine environment; or if copper and steel tubing are joined in a domestic water heater, the steel will corrode in the vicinity of the junction. Depending on the nature of the solution, one or more of reduction reactions will occur at the surface of the cathode material.

The rate of galvanic attack depends on the relative anode-to-cathode surface areas that are exposed to the electrolyte, and the rate is related directly to the cathode-anode area ratio; that is, for a given cathode area, a smaller anode will corrode more rapidly than a larger one.

A number of measures may be taken to significantly reduce the effects of galvanic corrosion. These include the following:
1. If coupling of dissimilar metals is necessary, choose two that are close together in the galvanic series.
2. Avoid an unfavorable anode-to-cathode surface area ratio; use an anode area as large as possible.
3. Electrically insulate dissimilar metals from each other.
4. Electrically connect a third, anodic metal to the other two; this is a form of cathodic protection.

Crevice corrosion

Electrochemical corrosion may also occur as a consequence of concentration differences of ions or dissolved gases in the electrolyte solution, and between two regions of the same metal piece. For such a concentration cell, corrosion occurs in the locale that has the lower concentration. A good example of this type of corrosion occurs in crevices and recesses or under deposits of dirt or corrosion products where the solution becomes stagnant and there is localized depletion of dissolved oxygen. Corrosion preferentially occurring at these positions is called crevice corrosion (Figure 7.1). The crevice must be wide enough for the solution to penetrate, yet narrow enough for stagnancy; usually the width is several thousandths of an inch.

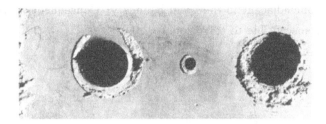

Figure 7.1 On this plate, which was immersed in seawater, crevice corrosion has occurred at the regions that were covered by washers

Crevice corrosion may be prevented by using welded instead of riveted or bolted joints, using nonabsorbing gaskets when possible, removing accumulated deposits frequently, and designing containment vessels to avoid stagnant areas and ensure complete drainage.

Pitting

Pitting is another form of very localized corrosion attack in which small pits or holes form. They ordinarily penetrate from the top of a horizontal surface downward in a nearly vertical direction. It is an extremely insidious type of corrosion, often going undetected and with very little material loss until failure occurs. An example of pitting corrosion is shown in Figure 7.2.

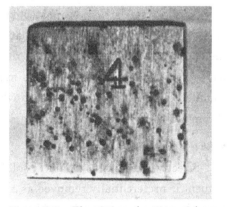

The mechanism for pitting is probably the same as for crevice corrosion in that oxidation occurs within the pit itself, with complementary reduction at the surface. It is supposed that gravity causes the pits to grow downward, the solution at the pit tip becoming more concentrated and dense as pit growth progresses. A pit may be initiated by a localized surface defect such as a scratch or a slight

Figure 7.2 The pitting of a 304 stainless steel plate by an acid-chloride solution

variation in composition. In fact, it has been observed that specimens having polished surfaces display a greater resistance to pitting corrosion. Stainless steels are somewhat susceptible to this form of corrosion; however, alloying with about 2% molybdenum enhances their resistance significantly.

Intergranular corrosion

As the name suggests, intergranular corrosion occurs preferentially along grain boundaries for some alloys and in specific environments. The net result is that a macroscopic specimen disintegrates along its grain boundaries. This type of corrosion is especially prevalent in some stainless steels. When heated to temperatures between 500 and 800℃ for sufficiently long time periods, these alloys become sensitized to intergranular attack. It is believed that this heat treatment permits the formation of small precipitate particles of chromium carbide ($Cr_{23}C_6$) by reaction between the chromium and carbon in the stainless steel. These particles form along the grain boundaries, as illustrated in Figure 7.3. Both the chromium and the carbon must diffuse to the grain boundaries to form the precipitates, which leaves a chromium-depleted zone adjacent to the grain boundary. Consequently, this grain boundary region is now highly susceptible to corrosion.

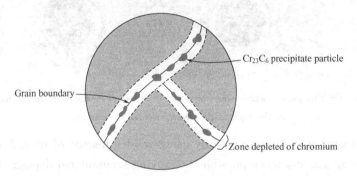

Figure 7.3 Schematic illustration of chromium carbide particles that have precipitated along grain boundaries in stainless steel, and the attendant zones of chromium depletion

Intergranular corrosion is an especially severe problem in the welding of stainless steels, when it is often termed **weld decay**.

Stainless steels may be protected from intergranular corrosion by the following measures: ①subjecting the sensitized material to a high-temperature heat treatment in which all the chromium carbide particles are redissolved, ②lowering the carbon content below 0.03 wt% C so that carbide formation is minimal, and③alloying the stainless steel with another metal such as niobium or titanium, which has a greater tendency to form carbides than does chromium so that the Cr remains in solid solution.

Selective leaching

Selective leaching is found in solid solution alloys and occurs when one element or constituent is preferentially removed as a consequence of corrosion processes. The most common example is the dezincification of brass, in which zinc is selectively leached from a copper-zinc brass alloy. The mechanical properties of the alloy are significantly impaired, since only a porous mass of copper remains in the region that has been dezincified. In addition, the material changes from yellow to a red or copper color. Selective leaching may also occur with other alloy systems in which aluminum, iron, cobalt, chromium, and other elements are vulnerable to preferential removal.

Erosion-corrosion

Erosion-corrosion arises from the combined action of chemical attack and mechanical abrasion or wear as a consequence of fluid motion. Virtually all metal alloys, to one degree or another, are susceptible to erosion-corrosion. It is especially harmful to alloys that passivate by forming a protective surface film; the abrasive action may erode away the film, leaving exposed a bare metal surface. If the coating is not capable of continuously and rapidly reforming as a protective barrier, corrosion may be severe. Relatively soft metals such as copper and lead are also sensitive to this form of attack. Usually it can be identified by surface grooves and waves having contours that are characteristic of the flow of the fluid.

Stress corrosion

Stress corrosion, sometimes termed stress corrosion cracking, results from the combined action of an applied tensile stress and a corrosive environment; both influences are necessary. In fact, some materials that are virtually inert in a particular corrosive medium become susceptible to this form of corrosion when a stress is applied. Small cracks form and then propagate in a direction perpendicular to the stress (Figure 7.4), with the result that failure may eventually occur. Failure behavior is characteristic of that for a brittle material, even though the metal alloy is intrinsically ductile. Furthermore, cracks may form at relatively low stress levels, significantly below the tensile strength. Most alloys are susceptible to stress corrosion in specific environments, especially at moderate stress levels. For example, most stressed stainless steels corrode in solutions containing chloride ions, whereas brasses are especially vulnerable when exposed to ammonia.

The stress that produces stress corrosion cracking need not be externally applied; it may be a residual one that results from rapid temperature changes and uneven contraction, or for two-phase alloys in which each phase has a different coefficient of expansion. Also, gaseous and solid corrosion products that are entrapped internally can give rise to internal stresses.

Probably the best measure to take in reducing or totally eliminating stress corrosion is to lower the magnitude of the stress. This may be accomplished by reducing the external load or increasing the cross-sectional area perpendicular to the applied stress. Furthermore, an appropriate heat treatment may be used to anneal out any residual thermal stresses.

Figure 7.4 A bar of steel that has been bent into a "horseshoe" shape using a nut-and-bolt assembly. While immersed in seawater, stress corrosion cracks formed along the bend at those regions where the tensile stresses are the greatest

Hydrogen Embrittlement

Various metal alloys, specifically some steels, experience a significant reduction in ductility and tensile strength when atomic hydrogen (H) penetrates into the material. This phenomenon is aptly referred to as hydrogen embrittlement. Strictly speaking, hydrogen embrittlement is a type of failure; in response to applied or residual tensile stresses, brittle fracture occurs catastrophically as cracks grow and rapidly propagate. Hydrogen in its atomic form (H as opposed to the molecular form, H_2) diffuses interstitially through the crystal lattice, and concentrations as low as several parts per million can lead to cracking. Furthermore, hydrogen induced cracks are most often transgranular, although intergranular fracture is observed for some alloy systems. A number of mechanisms have been proposed to explain hydrogen embrittlement; most of them are based on the interference of dislocation motion by the dissolved hydrogen.

For hydrogen embrittlement to occur, some source of hydrogen must be present, and, in addition, the possibility for the formation of its atomic species. Situations wherein these conditions are met include the following: pickling of steels (*Pickling* is a procedure used to remove surface oxide scale from steel pieces by dipping them in a vat of hot, dilute sulfuric or hydrochloric acid. in sulfuric acid); electroplating; and the presence of hydrogen-bearing atmospheres (including water vapor) at elevated temperatures such as during welding and heat treatments. Also, the presence of what are termed "poisons" such as sulfur (i. e., H_2S) and arsenic compounds accelerates hydrogen embrittlement; these substances retard the formation of molecular hydrogen and thereby increase the residence time of atomic hydrogen on the metal surface. Hydrogen sulfide, probably the most aggressive poison, is found in petroleum fluids, natural gas, oil-well brines, and geothermal fluids.

Some of the techniques commonly used to reduce the likelihood of hydrogen embrittlement include reducing the tensile strength of the alloy via a heat treatment, removal of the source of hydrogen, "baking" the alloy at an elevated temperature to drive out any dissolved hydrogen, and substitution of a more embrittlement resistant alloy.

7.2.6 Corrosion Environments

Corrosive environments include the atmosphere, aqueous solutions, soils, acids, bases, inorganic solvents, molten salts, liquid metals, and, last but not least, the human body. On a tonnage basis, atmospheric corrosion accounts for the greatest losses. Moisture containing dissolved oxygen is the primary corrosive agent, but other substances, including sulfur compounds and sodium chloride, may also contribute. This is especially true of marine atmospheres, which are highly corrosive because of the presence of sodium chloride. Dilute sulfuric acid solutions (acid rain) in industrial environments can also cause corrosion problems. Metals commonly used for atmospheric applications include alloys of aluminum and copper, and galvanized steel.

Water environments can also have a variety of compositions and corrosion characteristics. Freshwater normally contains dissolved oxygen, as well as other minerals several of which account for hardness. Seawater contains approximately 3.5% salt (predominantly sodium chloride), as well as some minerals and organic matter. Seawater is generally more corrosive than freshwater, frequently producing pitting and crevice corrosion. Cast iron, steel, aluminum, copper, brass, and some stainless steels are generally suitable for freshwater use, whereas titanium, brass, some bronzes, copper-nickel alloys, and nickel-chromium-molybdenum alloys are highly corrosion resistant in seawater.

Soils have a wide range of compositions and susceptibilities to corrosion. Compositional variables include moisture, oxygen, salt content, alkalinity, and acidity, as well as the presence of various forms of bacteria. Cast iron and plain carbon steels, both with and without protective surface coatings, are found most economical for underground structures.

7.2.7 Corrosion Prevention

Some corrosion prevention methods were treated relative to the eight forms of corrosion; however, only the measures specific to each of the various corrosion types were discussed.

Physical barriers to corrosion, as a general technique, are applied on surfaces in the form of films and coatings. A large diversity of metallic and nonmetallic coating materials are available. It is essential that the coating maintain a high degree of surface adhesion, which undoubtedly requires some preapplication surface treatment. In most cases, the coating must be virtually nonreactive in the corrosive environment and resistant to mechanical damage that exposes the bare metal to the corrosive environment. All three material types—metals, ceramics, and polymers—are used as coatings for metals.

Another effective method of corrosion prevention is cathodic protection; it can be used for all eight different forms of corrosion as discussed above, and may, in some situations, completely stop corrosion. One cathodic protection technique employs a galvanic couple: the metal to be protected is electrically connected to another metal that is more reactive in the particular environment. The latter experiences oxidation, and, upon giving up electrons, protects the first metal from corrosion. The oxidized metal is often called a sacrificial anode, and magnesium and zinc are commonly used as such because they lie at the anodic end of the galvanic series. The process of galvanizing is simply one in which a layer of zinc is applied to the surface of steel by hot dipping. In the atmosphere and most aqueous environments, zinc is anodic to and will thus cathodically protect the steel if there is any surface damage. Any corrosion of the zinc coating will proceed at an extremely slow rate because the ratio of the anode-to-cathode surface area is quite large.

7.3 Corrosion of Ceramic Materials

Ceramic materials, being compounds between metallic and nonmetallic elements, may be thought of as having already been corroded. Thus, they are exceedingly immune to corrosion by almost all environments, especially at room temperature. Corrosion of ceramic materials generally involves simple chemical dissolution, in contrast to the electrochemical processes found in metals, as described above.

Ceramic materials are frequently utilized because of their resistance to corrosion. Glass is often used to contain liquids for this reason. Refractory ceramics must not only withstand high temperatures and provide thermal insulation but, in many instances, must also resist high-temperature attack by molten metals, salts, slags, and glasses. Some of the new technology schemes for converting energy from one form to another that is more useful require relatively high temperatures, corrosive atmospheres, and pressures above the ambient. Ceramic materials are much better suited to withstand most of these environments for reasonable time periods than are metals.

7.4 Degradation of Polymers

Polymeric materials also experience deterioration by means of environmental interactions. However, an undesirable interaction is specified as degradation rather than corrosion because the processes are basically dissimilar. Whereas most metallic corrosion reactions are electrochemical, by contrast, polymeric degradation is physiochemical; that is, it involves physical as well as chemical phenomena. Furthermore, a wide variety of reactions and adverse consequences are possible for polymer degradation. Polymers may deteriorate by swelling and dissolution. Covalent bond rupture, as a result of heat energy, chemical reactions, and radiation is also possible, ordinarily with an attendant reduction in mechanical integrity. It should also be mentioned that because of the chemical complexity of polymers, their degradation mechanisms are not well understood.

7.4.1 Swelling and Dissolution

When polymers are exposed to liquids, the main forms of degradation are swelling and dissolution. With swelling, the liquid or solute diffuses into and is absorbed within the polymer; the small solute molecules fit into and occupy positions among the polymer molecules. Thus the macromolecules are forced apart such that the specimen expands or swells. Furthermore, this increase in chain separation results in a reduction of the secondary intermolecular bonding forces; as a consequence, the material becomes softer and more ductile. The liquid solute also lowers the glass transition temperature and, if depressed below the ambient temperature, will cause a once strong material to become rubbery and weak.

Swelling may be considered to be a partial dissolution process in which there is only limited solubility of the polymer in the solvent. Dissolution, which occurs when the polymer is completely soluble, may be thought of as just a continuation of swelling. For example, many hydrocarbon rubbers readily absorb hydrocarbon liquids such as gasoline.

In general, increasing molecular weight, increasing degree of crosslinking and crystallinity, and decreasing temperature result in a reduction of these deteriorative processes.

7.4.2 Bond Rupture

Polymers may also experience degradation by a process termed **scission**—the severance or rupture of molecular chain bonds. This causes a separation of chain segments at the point of scission and a reduction in the molecular weight. Bond rupture may result from exposure to radiation or to heat, and from chemical reaction.

Certain types of radiation, like electron beams, X-rays, β- and γ-rays, and ultraviolet (UV) radiation, possess sufficient energy to penetrate a polymer specimen and interact with the constituent atoms or their electrons. One such reaction is **ionization**, in which the radiation removes an orbital electron from a specific atom, converting that atom into a positively charged ion. As a consequence, one of the covalent bonds associated with the specific atom is broken, and there is a rearrangement of atoms or groups of atoms at that point. This bond breaking leads to either scission or crosslinking at the ionization site, depending on the chemical structure of the polymer and also on the dose of radiation. In day-to-day use, the greatest radiation damage to polymers is caused by UV irradiation. After prolonged exposure, most polymer films become brittle, discolor, crack, and fail. For example, camping tents begin to tear, dashboards develop cracks, and plastic windows become cloudy.

Oxygen, ozone, and other substances can cause or accelerate chain scission as a result of chemical reaction. This effect is especially prevalent in vulcanized rubbers that have doubly bonded carbon atoms along the backbone molecular chains, and that are exposed to ozone (O_3), an atmospheric pollutant. Ordinarily, if the rubber is in an unstressed state, a film will form on the surface, protecting the bulk material from any further reaction. However, when these materials are subjected to tensile stresses, cracks and crevices form and grow in a direction perpendicular to the stress; eventually, rupture of the material may occur. This is why the sidewalls on rubber bicycle tires develop cracks as they age. Apparently these cracks result from large numbers of ozone-induced scissions. Chemical degradation is a particular problem for polymers used in areas with high levels of air pollutants such as smog and ozone.

Thermal degradation corresponds to the scission of molecular chains at elevated temperatures; as a consequence, some polymers undergo chemical reactions in which gaseous species are produced. These reactions are evidenced by a weight loss of material; a polymer's thermal stability is a measure of its resilience to this decomposition. Thermal stability is related primarily to the magnitude of the bonding energies between the various atomic constituents of the polymer; higher bonding energies result in more thermally stable materials.

7.4.3 Weathering

Many polymeric materials serve in applications that require exposure to outdoor conditions. Any resultant degradation is termed weathering, which may, in fact, be a combination of several different processes. Under these conditions deterioration is primarily a result of oxidation, which is initiated by ultraviolet radiation from the sun. Some polymers such as nylon and cellulose are also susceptible to water absorption, which produces a reduction in their hardness and stiffness. Resistance to weathering among the various polymers is quite diverse. The fluorocarbons are virtually inert under these conditions; but some materials, including poly (vinyl chloride) and polystyrene, are susceptible to weathering.

Summary

Electrochemical Considerations

Metallic corrosion is ordinarily electrochemical, involving both oxidation and reduction reactions. Oxidation is the loss of the metal atom's valence electrons; the resulting metal ions may either go into the corroding solution or form an insoluble compound. During reduction, these electrons are transferred to at least one other chemical species. The character of the corrosion environment dictates which of several possible reduction reactions will occur.

Not all metals oxidize with the same degree of ease, which is demonstrated with a galvanic couple; when in an electrolyte, one metal (the anode) will corrode, whereas a reduction reaction will occur at the other metal (the cathode). The magnitude of the electric potential that is established between anode and cathode is indicative of the driving force for the corrosion reaction.

The standard emf and galvanic series are simply rankings of metallic materials on the basis of their tendency to corrode when coupled to other metals. For the standard emf series, ranking is based on the magnitude of the voltage generated when the standard cell of a metal is coupled to the standard hydrogen electrode at 25℃. The galvanic series consists of the relative reactivities of metals and alloys in seawater.

The half-cell potentials in the standard emf series are thermodynamic parameters that are valid only at equilibrium; corroding systems are not in equilibrium. Furthermore, the magnitudes of these potentials provide no indication as to the rates at which corrosion reactions occur.

Corrosion Rates

The rate of corrosion may be expressed as corrosion penetration rate, that is, the thickness loss of material per unit of time. Mils per year and millimeters per year are the common units for this parameter. Alternatively, rate is proportional to the current density associated with the electrochemical reaction.

Passivity

A number of metals and alloys passivate, or lose their chemical reactivity, under some environmental circumstances. This phenomenon is thought to involve the formation of a thin protective oxide film. Stainless steels and aluminum alloys exhibit this type of behavior. The active-to-passive behavior may be explained by the alloy's S-shaped electrochemical potential-versus-log current density curve. Intersections with reduction polarization curves in active and passive regions correspond, respectively, to high and low corrosion rates.

Forms of Corrosion

Metallic corrosion is sometimes classified into eight different forms: uniform attack, galvanic corrosion, crevice corrosion, pitting, intergranular corrosion, selective leaching, erosion-corrosion, and stress corrosion. Hydrogen embrittlement, a type of failure sometimes observed in corrosion environments, was also discussed.

Corrosion Prevention

The measures that may be taken to prevent, or at least reduce, corrosion include material selection, environmental alteration, the use of inhibitors, design changes, application of coatings, and cathodic protection.

Corrosion of Ceramic Materials

Ceramic materials, being inherently corrosion resistant, are frequently utilized at elevated temperatures and/or in extremely corrosive environments.

Swelling and Dissolution

Bond Rupture

Weathering

Polymeric materials deteriorate by noncorrosive processes. Upon exposure to liquids, they may experience degradation by swelling or dissolution. With swelling, solute molecules actually fit into the molecular structure. Scission, or the severance of molecular chain bonds, may be induced by radiation, chemical reactions, or heat. This results in a reduction of molecular weight and a deterioration of the physical and chemical properties of the polymer.

Important Terms and Concepts

Activation polarization	Anode	Cathode
Cathodic protection	Concentration polarization	Corrosion
Corrosion penetration rate	Crevice corrosion	Degradation
Electrolyte	Electromotive force(emf) series	Erosion-corrosion
Galvanic corrosion	Galvanic series	Hydrogen embrittlement
Inhibitor	Intergranular corrosion	Molarity
Oxidation	Passivity	Pilling-Bedworth ratio
Pitting	Polarization	Reduction
Sacrificial anode	Scission	Selective leaching
Standard half-cell	Stress corrosion	Weld decay

References

[1] *ASM Handbook*, Vol. 13, *Corrosion*, ASM International, Materials Park, OH, 1987.
[2] *ASM Handbook*, Vol. 13A, *Corrosion: Fundamentals, Testing, and Protection*, ASM International, Materials Park, OH, 2003.
[3] Craig, B. D. and D. Anderson, (Editors), *Handbook of Corrosion Data*, 2nd edition, ASM International, Materials Park, OH, 1995.
[4] Fontana, M. G., *Corrosion Engineering*, 3rd edition, McGraw-Hill, New York, 1986.
[5] Gibala, R. and R. F. Hehemann, *Hydrogen Embrittlement and Stress Corrosion Cracking*, ASM International, Materials Park, OH, 1984.
[6] Jones, D. A., *Principles and Prevention of Corrosion*, 2nd edition, Pearson Education, Upper Saddle River, NJ, 1996.
[7] Marcus, P. and J. Oudar (Editors), *Corrosion Mechanisms in Theory and Practice*, Marcel Dekker, New York, 1995.
[8] Revie, R. W. (Editor), *Uhlig's Corrosion Handbook*, 2nd edition, John Wiley & Sons, New York, 2000.
[9] Schweitzer, P. A., *Atmospheric Degradation and Corrosion Control*, Marcel Dekker, New York, 1999.
[10] Schweitzer, P. A. (Editor), *Corrosion and Corrosion Protection Handbook*, 2nd edition, Marcel Dekker, New York, 1989.
[11] Talbot, D. and J. Talbot, *Corrosion Science and Technology*, CRC Press, Boca Raton, FL, 1998.
[12] Uhlig, H. H. and R. W. Revie, *Corrosion and Corrosion Control*, 3rd edition, John Wiley & Sons, New York, 1985.

Chapter 8 Electrical/Thermal/Magnetic/ Optical Properties of Materials

Learning Objectives
After careful study of this chapter you should be able to do the following:

Electrical Properties of Materials
1. Describe the four possible electron band structures for solid materials.
2. Briefly describe electron excitation events that produce free electrons/holes in (a) metals, (b) semiconductors (intrinsic and extrinsic), and (c) insulators.
3. Calculate the electrical conductivities of metals, semiconductors (intrinsic and extrinsic), and insulators given their charge carrier density (s) and mobility (s).
4. Distinguish between intrinsic and extrinsic semiconducting materials.

Thermal Properties of Materials
5. Define heat capacity and specific heat.
6. Note the primary mechanism by which thermal energy is assimilated in solid materials.
7. Determine the linear coefficient of thermal expansion given the length alteration that accompanies a specified temperature change.
8. Briefly explain the phenomenon of thermal expansion from an atomic perspective using a potential energy-versus-interatomic separation plot.
9. Define thermal conductivity.

Magnetic Properties of Materials
10. Determine the magnetization of some material given its magnetic susceptibility and the applied magnetic field strength.
11. From an electronic perspective note and briefly explain the two sources of magnetic moments in materials.
12. Briefly explain the nature and source of (a) diamagnetism, (b) paramagnetism, and (c) ferromagnetism.
13. In terms of crystal structure, explain the source of ferrimagnetism for cubic ferrites.
14. (a) Describe magnetic hysteresis; (b) explain why ferromagnetic and ferrimagnetic materials experience magnetic hysteresis; and (c) explain why these materials may become permanent magnets.
15. Note the distinctive magnetic characteristics for both soft and hard magnetic materials.
16. Describe the phenomenon of superconductivity.

Optical Properties of Materials
17. Briefly describe electronic polarization that results from electromagnetic radiation-atomic interactions. Cite two consequences of electronic polarization.
18. Briefly explain why metallic materials are opaque to visible light.
19. Define index of refraction.

20. *Describe the mechanism of photon absorption for* (a) *high-purity insulators and semiconductors, and* (b) *insulators and semiconductors that contain electrically active defects.*

8.1 Introduction

Stone Age—Bronze Age—Iron Age—what's next? Some individuals have called the present era the *space age* or the *atomic age*. However, space exploration and nuclear reactors, to mention only two major examples, have only little impact on our everyday lives. Instead, electrical and electronic devices (such as radio, television, telephone, refrigerator, computers, electric light, CD players, electromotors, etc.) permeate our daily life to a large extent. Life without electronics would be nearly unthinkable in many parts of the world. The present era could, therefore, be called the age of electricity. However, electricity needs a medium in which to manifest itself and to be placed in service. For this reason, and because previous eras have been named after the material that had the largest impact on the lives of mankind, the present time may best be characterized by the name.

We are almost constantly in contact with electronic materials, such as conductors, insulators, semiconductors, (ferro) magnetic materials, optically transparent matter, and opaque substances. The useful properties of these materials are governed and are characterized by *electrons*. In fact, the terms *electronic materials* and *electronic properties* should be understood in the widest possible sense, meaning to include all phenomena in which electrons participate in an active (dynamic) role. This is certainly the case for electrical, thermal, magnetic, and even many optical phenomena.
In contrast to this, mechanical properties can be mainly interpreted by taking the interactions of *atoms* into account, as explained in previous chapters.

8.2 Electrical Properties of Materials

The prime objective of this section is to explore the electrical properties of materials, that is, their responses to an applied electric field. We begin with the phenomenon of electrical conduction: the parameters by which it is expressed, the mechanism of conduction by electrons, and how the electron energy band structure of a material influences its ability to conduct. These principles are extended to metals, semiconductors, and insulators. Particular attention is given to the characteristics of semiconductors. Also treated are the dielectric characteristics of insulating materials. The final sections are devoted to the peculiar phenomena of ferroelectricity and piezoelectricity.

One of the principal characteristics of materials is their ability (or lack of ability) to conduct electrical current. Indeed, materials are classified by this property, that is, they are divided into conductors, semiconductors, and nonconductors. (The latter are often called insulators or dielectrics.) The **conductivity**, σ, of different materials at room temperature spans more than 25 orders of magnitude, as depicted in Figure 8.1. Moreover, if one takes the conductivity of superconductors, measured at low temperatures, into consideration, this span extends to 40 orders of magnitude [using an estimated conductivity for superconductors of about 10^{20} $1/(\Omega \cdot cm)$]. This is the largest known variation in a physical property and is only comparable to the ratio between the diameter of the universe (about 10^{26} m) and the radius of an electron (10^{-14} m).

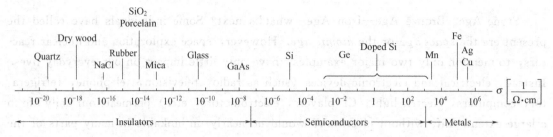

Figure 8.1 Room-temperature conductivity of various materials
(Superconductors, having conductivities of many orders of magnitude
larger than copper, near 0 K, are not shown. The conductivity of
semiconductors varies substantially with temperature and purity)

8.2.1 Metals and Alloys

We know that the electrons of isolated atoms (for example in a gas) can be considered to orbit at various distances about their nuclei. These orbits constitute different energies. Specifically, the larger the radius of an orbit, the larger the excitation energy of the electron. This fact is often represented in a somewhat different fashion by stating that the electrons are distributed on different *energy levels*, as schematically shown on the right side of Figure 8.2. Now, these distinct energy levels, which are characteristic for isolated atoms, widen into **energy bands** when atoms approach each other and eventually form a solid as depicted on the left side of Figure 8.2.

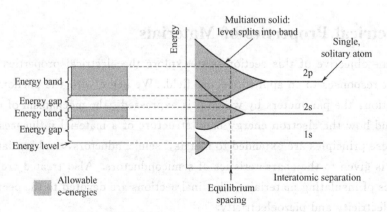

Figure 8.2 Schematic representation of energy levels
(as for isolated atoms) and widening of these levels into
energy bands with decreasing distance between atoms.
Energy bands for a specific case are shown at the left of the diagram

The number of states within each band will equal the total of all states contributed by the N atoms. For example, an s band will consist of N states, and a p band of $3N$ states. With regard to occupancy, each energy state may accommodate two electrons, which must have oppositely directed spins. Furthermore, bands will contain the electrons that resided in the corresponding levels of the isolated atoms; for example, a $4s$ energy band in the solid will contain those isolated atom's $4s$ electrons. Of course, there will be empty bands and, possibly, bands that are only partially filled.

The electrical properties of a solid material are a consequence of its electron band structure—that is, the arrangement of the outermost electron bands and the way in which they are filled with electrons. Notably energy bands and their occupation by electrons were introduced. It was suggested that metals were distinct from insulators, and that semiconductors occupied a middle ground between the two. Simple band diagram representations in Figure 8.3 graphically distinguish these three types of solids. Semiconductors have a relatively small energy gap, E_g, of ~1 eV separating the nearly full valence and almost empty conduction bands. Insulators have a larger energy gap (5~10eV) between the even fuller valence band and virtually unoccupied conduction band. The size of the energy gap may be viewed as a barrier to electrical conduction and that is why insulators are poor conductors. Conversely, metals have no effective energy gap because valence and conduction bands either overlap [Figure 8.3(a)] or the conduction band is only partially filled. Therefore, they are good conductors.

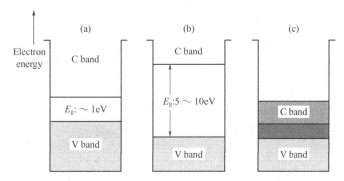

Figure 8.3 Comparison of electron band diagrams for (a) metals, (b) insulators, and (c) semiconductors

We may now return to the conductivity. In short, according to quantum theory, only those materials that possess partially filled electron bands are capable of conducting an electric current. Electrons can then be lifted slightly above the Fermi energy into an allowed and unfilled energy state. This permits them to be accelerated by an electric field, thus producing a current. Second, only those electrons that are close to the Fermi energy participate in the electric conduction. (The classical electron theory taught us instead that *all* free electrons would contribute to the current.) Third, the number of electrons near the Fermi energy depends on the density of available electron states.

8.2.2 Semiconductors

Semiconductors such as silicon or germanium are neither good conductors nor good insulators as seen in Figure 8.1. This may seem to make semiconductors to be of little interest.

Their usefulness results, however, from a completely different property, namely, that extremely small amounts of certain impurity elements, which are called *dopants*, remarkably change the electrical behavior of semiconductors. Indeed, semiconductors have been proven in recent years to be the lifeblood of a multibillion dollar industry which prospers essentially from this very feature. Silicon, the major species of semiconducting materials, is today the single most researched element. Silicon is abundant (28% of the earth's crust consists of it); the raw material (SiO_2 or sand) is inexpensive; Si forms a natural, insulating oxide; its heat conduction is reasonable; it is nontoxic; and it is stable against environmental influences.

Intrinsic Semiconductors

The properties of semiconductors are commonly explained by making use of the already introduced electron band structure which is the result of quantum-mechanical considerations. In simple terms, the electrons are depicted to reside in certain allowed energy regions. Specifically, Figure 8.3(c) and Figure 8.4 depict two electron bands, the lower of which, at 0 K, is completely filled with valence electrons. This band is appropriately called the **valence band**. It is separated by a small gap (about 1.1 eV for Si) from the **conduction band**, which, at 0 K, contains no electrons. Further, quantum mechanics stipulates that electrons essentially are not allowed to reside in the gap between these bands (called the *forbidden band*). Since the filled valence band possesses no allowed empty energy states in which the electrons can be thermally excited (and then accelerated in an electric field), and since the conduction band contains no electrons at all, silicon, at 0 K, is an insulator.

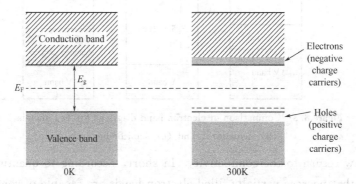

Figure 8.4 Simplified band diagrams for an intrinsic semiconductor such as pure silicon at two different temperatures. The dark shading symbolizes electrons

Extrinsic Semiconductors

The number of electrons in the conduction band can be considerably increased by adding, for example, to silicon small amounts of elements from Group V of the Periodic Table called *donor atoms*. Dopants such as phosphorous or arsenic are commonly utilized, which are added in amounts of, for example, 0.0001%. These dopants replace some regular lattice atoms in a substitutional manner. Since phosphorous has five valence electrons, that is, one more than silicon, the extra electron, called the *donor electron*, is only loosely bound. The binding energy of phosphorous donor electrons in a silicon matrix, for example, is about 0.045 eV. Thus, the donor electrons can be disassociated from their nuclei by only a slight increase in thermal energy. Indeed, at room temperature all donor electrons have already been excited into the conduction band.

It is common to describe this situation by introducing into the forbidden band so-called *donor levels*, which accommodate the donor electrons at 0 K; see Figure 8.5 (a). The distance between the donor level and the conduction band represents the energy that is needed to transfer the extra electrons into the conduction band (e.g., 0.045 eV for P in Si). The electrons that have been excited from the donor levels into the conduction band are free and can be accelerated in an electric field as shown in Figure 8.2 and Figure 8.4. Since the conduction mechanism in semiconductors with donor impurities is predominated by *negative* charge carriers, one calls these materials *n-type semiconductors*. Similar considerations may be carried out with respect to impurities from the third group of the Periodic Table (B, Al, Ga, In). They are deficient in one electron compared to silicon and therefore tend to accept an electron. The conduction mechanism in these semiconductors with *acceptor impurities* is thus predominated by *positive* charge carriers (holes) which are introduced from the *acceptor levels* [Figure 8.5(b)] into the valence band. They are therefore called *p-type* semiconductors. In other words, the conduction in p-type semiconductors under the influence of an external electric field occurs in the valence band and is predominated by holes.

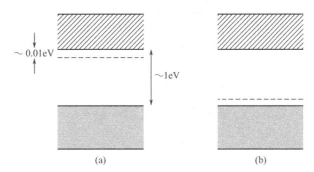

Figure 8.5 (a) Donor and (b) acceptor levels in extrinsic semiconductors

Compound Semiconductors

Compounds made of Group Ⅲ and Group Ⅴ elements, such as gallium arsenide, have similar semiconducting properties as the Group Ⅳ materials silicon or germanium. GaAs is of some technical interest because of its above-mentioned wider band gap, and because of its larger electron mobility, which aids in highspeed applications. Further, the ionization energies of donor and acceptor impurities in GaAs are one order of magnitude smaller than in silicon, which ensures complete electron (and hole) transfer from the donor (acceptor) levels into the conduction (valence) bands, even at relatively low temperatures. However, GaAs is about ten times more expensive than Si and its heat conduction is smaller. Other compound semiconductors include Ⅱ-Ⅵ combinations such as ZnO, ZnS, ZnSe, or CdTe, and Ⅳ-Ⅵ materials such as PbS, PbSe, or PbTe. Silicon carbide, a Ⅳ-Ⅳ compound, has a band gap of 3 eV and can thus be used up to 700℃ before intrinsic effects set in. The most important application of compound semiconductors is, however, for opto-electronic purposes (e.g.

for light-emitting diodes, and lasers).

8.2.3 Ionic Ceramics and Polymers

Most polymers and ionic ceramics are insulating materials at room temperature and, therefore, have electron energy band structures similar to that represented in Figure 8.3(c); a filled valence band is separated from an empty conduction band by a relatively large band gap, usually greater than 2 eV. Thus, at normal temperatures only very few electrons may be excited across the band gap by the available thermal energy, which accounts for the very small values of conductivity; Table 8.1 gives the room- temperature electrical conductivities of several of these materials. Of course, many materials are utilized on the basis of their ability to insulate, and thus a high electrical resistivity is desirable. With rising temperature, insulating materials experience an increase in electrical conductivity, which may ultimately be greater than that for semiconductors.

Table 8.1 Typical Room-Temperature Electrical Conductivities for 13 Nonmetallic Materials

Material	Electrical Conductivity/$[(\Omega \cdot m)^{-1}]$
Graphite	$3 \times 10^4 \sim 2 \times 10^5$
Ceramics	
Concrets(dry)	10^{-9}
Soda-lime glass	$10^{-10} \sim 10^{-11}$
Porcelain	$10^{-10} \sim 10^{-12}$
Borosilicate glass	$\sim 10^{-13}$
Aluminum oxide	$< 10^{-13}$
Fused silica	$< 10^{-18}$
Polymers	
Phenol-formaldehyde	$10^{-9} \sim 10^{-10}$
Poly(methyl methacrylate)	$< 10^{-12}$
Nylon 6,6	$10^{-12} \sim 10^{-13}$
Polystyrene	$< 10^{-14}$
Polyethylene	$10^{-15} \sim 10^{-17}$
Polytetrafluoroethylene	$< 10^{-17}$

8.3 Thermal Properties of Materials

By "thermal property" is meant the response of a material to the application of heat. As a solid absorbs energy in the form of heat, its temperature rises and its dimensions increase. The energy may be transported to cooler regions of the specimen if temperature gradients exist, and ultimately, the specimen may melt. Heat capacity, thermal expansion, and thermal conductivity are properties that are often critical in the practical utilization of solids.

8.3.1 Heat Capacity

A solid material, when heated, experiences an increase in temperature signifying that some energy has been absorbed. **Heat capacity** is a property that is indicative of a material's ability to absorb heat from the external surroundings; it represents the amount of energy required to produce a unit temperature rise. In mathematical terms, the heat capacity C is expressed as follows:

$$C = \frac{dQ}{dT} \tag{8.1}$$

where dQ is the energy required to produce a dT temperature change. Ordinarily, heat capacity is specified per mole of material (e. g., J/mol-K, or cal/mol-K). **Specific heat** (often denoted by a lowercase c) is sometimes used; this represents the heat capacity per unit mass and has various units (J/kg-K, cal/g-K, Btu/lbm-°F).

There are really two ways in which this property may be measured, according to the environmental conditions accompanying the transfer of heat. One is the heat capacity while maintaining the specimen volume constant, C_v; the other is for constant external pressure, which is denoted C_p. The magnitude of C_p is almost always greater than C_v; however, this difference is very slight for most solid materials at room temperature and below.

8.3.2 Thermal Expansion

Most solid materials expand upon heating and contract when cooled. The change in length with temperature for a solid material may be expressed as follows:

$$\frac{l_f - l_0}{l_0} = \alpha_l (T_f - T_0) \tag{8.2a}$$

or

$$\frac{\Delta l}{l_0} = \alpha_l \Delta T \tag{8.2b}$$

where l_0 and l_f represent, respectively, initial and final lengths with the temperature change from to The T_0 to T_f. he parameter α_l is called the **linear coefficient of thermal expansion**; it is a material property that is indicative of the extent to which a material expands upon heating, and has units of reciprocal temperature [(℃)$^{-1}$ or (℉)$^{-1}$]. Of course, heating or cooling affects all the dimensions of a body, with a resultant change in volume. Volume changes with temperature may be computed from

$$\frac{\Delta V}{V} = \alpha_v \Delta T \tag{8.3}$$

where ΔV and V_0 are the volume change and the original volume, respectively, and α_v symbolizes the volume coefficient of thermal expansion. In many materials, the value of α_v is anisotropic; that is, it depends on the crystallographic direction along which it is measured. For materials in which the thermal expansion is isotropic, α_v is approximately 3 α_l.

Metals

Linear coefficients of thermal expansion for some of the common metals range between about 5×10^{-6} and 25×10^{-6} (℃)$^{-1}$; these values are intermediate in magnitude between those for ceramic and polymeric materials. As the following Materials of Importance piece explains, several low-expansion and controlled-expansion metal alloys have been developed, which are used in applications requiring dimensional stability with temperature variations.

Ceramics

Relatively strong interatomic bonding forces are found in many ceramic materials as reflected in comparatively low coefficients of thermal expansion; values typically range between about 0.5×10^{-6} and 15×10^{-6} $(\text{℃})^{-1}$. For noncrystalline ceramics and also those having cubic crystal structures, a_l is isotropic. Otherwise, it is anisotropic; and, in fact, some ceramic materials, upon heating, contract in some crystallographic directions while expanding in others. For inorganic glasses, the coefficient of expansion is dependent on composition. Fused silica (high-purity SiO_2 glass) has a small expansion coefficient, 0.4×10^{-6} $(\text{℃})^{-1}$. This is explained by a low atomic packing density such that interatomic expansion produces relatively small macroscopic dimensional changes.

Ceramic materials that are to be subjected to temperature changes must have coefficients of thermal expansion that are relatively low, and in addition, isotropic. Otherwise, these brittle materials may experience fracture as a consequence of nonuniform dimensional changes in what is termed **thermal shock**, as discussed later in the chapter.

Polymers

Some polymeric materials experience very large thermal expansions upon heating as indicated by coefficients that range from approximately 50×10^{-6} to 400×10^{-6} $(\text{℃})^{-1}$. The highest a_l values are found in linear and branched polymers because the secondary intermolecular bonds are weak, and there is a minimum of cross-linking. With increased crosslinking, the magnitude of the expansion coefficient diminishes; the lowest coefficients are found in the thermosetting network polymers such as phenol-formaldehyde, in which the bonding is almost entirely covalent.

8.3.3 Thermal Conductivity

Thermal conduction is the phenomenon by which heat is transported from high- to low-temperature regions of a substance. The property that characterizes the ability of a material to transfer heat is the **thermal conductivity**.

Mechanisms of Heat Conduction

Heat is transported in solid materials by both lattice vibration waves (phonons) and free electrons. Free or conducting electrons participate in electronic thermal conduction. To the free electrons in a hot region of the specimen is imparted a gain in kinetic energy. They then migrate to colder areas, where some of this kinetic energy is transferred to the atoms themselves (as vibrational energy) as a consequence of collisions with phonons or other imperfections in the crystal. The relative contribution of to the total thermal conductivity increases with increasing free electron concentrations, since more electrons are available to participate in this heat transference process.

8.3.4 Thermal Stresses

Thermal stresses are stresses induced in a body as a result of changes in temperature. An understanding of the origins and nature of thermal stresses is important because these stresses can lead to fracture or undesirable plastic deformation.

Stresses Resulting From Restrained Thermal Expansion and Contraction

Let us first consider a homogeneous and isotropic solid rod that is heated or cooled uniformly; that is, no temperature gradients are imposed. For free expansion or contraction, the rod will be stress free. If, however, axial motion of the rod is restrained by rigid end supports, thermal stresses will be introduced. The magnitude of the stress σ resulting from a temperature change from T_0 to T_f is

$$\sigma = E\alpha_l(T_0 - T_f) = E\alpha_l \Delta T \tag{8.4}$$

where E is the modulus of elasticity and α_l is the linear coefficient of thermal expansion. Upon heating ($T_f > T_0$), the stress is compressive ($\sigma < 0$) since rod expansion has been constrained. Of course, if the rod specimen is cooled ($T_f < T_0$), a tensile stress will be imposed ($\sigma > 0$). Also, the stress is the same as the stress that would be required to elastically compress (or elongate) the rod specimen back to its original length after it had been allowed to freely expand (or contract) with the $T_0 - T_f$ temperature change.

Stresses Resulting From Temperature Gradients

When a solid body is heated or cooled, the internal temperature distribution will depend on its size and shape, the thermal conductivity of the material, and the rate of temperature change. Thermal stresses may be established as a result of temperature gradients across a body, which are frequently caused by rapid heating or cooling, in that the outside changes temperature more rapidly than the interior; differential dimensional changes serve to restrain the free expansion or contraction of adjacent volume elements within the piece. For example, upon heating, the exterior of a specimen is hotter and, therefore, will have expanded more than the interior regions. Hence, compressive surface stresses are induced and are balanced by tensile interior stresses. The interior-exterior stress conditions are reversed for rapid cooling such that the surface is put into a state of tension.

Thermal Shock of Brittle Materials

For ductile metals and polymers, alleviation of thermally induced stresses may be accomplished by plastic deformation. However, the nonductility of most ceramics enhances the possibility of brittle fracture from these stresses. Rapid cooling of a brittle body is more likely to inflict such thermal shock than heating, since the induced surface stresses are tensile. Crack formation and propagation from surface flaws are more probable when an imposed stress is tensile.

The capacity of a material to withstand this kind of failure is termed its *thermal shock resistance*. For a ceramic body that is rapidly cooled, the resistance to thermal shock depends not only on the magnitude of the temperature change, but also on the mechanical and thermal properties of the material. The thermal shock resistance is best for ceramics that have high fracture strengths σ_f and high thermal conductivities, as well as low moduli of elasticity and low coefficients of thermal expansion.

The resistance of many materials to this type of failure may be approximated by a thermal shock resistance parameter *TSR*:

$$TSR \cong \frac{\sigma_f k}{E\alpha_l} \tag{8.5}$$

Thermal shock may be prevented by altering the external conditions to the degree that cooling or heating rates are reduced and temperature gradients across a body are minimized. Modification of the thermal and/or mechanical characteristics in Equation 8.9 may also enhance the thermal shock resistance of a material. Of these parameters, the coefficient of thermal expansion is probably most easily changed and controlled. For example, common soda-lime glasses, which have an α_l of approximately 9×10^{-6} (℃)$^{-1}$, are particularly susceptible to thermal shock, as anyone who has baked can probably attest. Reducing the CaO and Na$_2$O contents while at the same time adding B$_2$O$_3$ in sufficient quantities to form borosilicate (or Pyrex) glass will reduce the coefficient of expansion to about 3×10^{-6} (℃)$^{-1}$; this material is entirely suitable for kitchen oven heating and cooling cycles. The introduction of some relatively large pores or a ductile second phase may also improve the thermal shock characteristics of a material; both serve to impede the propagation of thermally induced cracks.

It is often necessary to remove thermal stresses in ceramic materials as a means of improving their mechanical strengths and optical characteristics. This may be accomplished by an annealing heat treatment, as discussed for glasses in Chapter 5.

8.4　Magnetic Properties of Materials

Magnetism, the phenomenon by which materials assert an attractive or repulsive force or influence on other materials, has been known for thousands of years. However, the underlying principles and mechanisms that explain the magnetic phenomenon are complex and subtle, and their understanding has eluded scientists until relatively recent times. Many of our modern technological devices rely on magnetism and magnetic materials; these include electrical power generators and transformers, electric motors, radio, television, telephones, computers, and components of sound and video reproduction systems.

Iron, some steels, and the naturally occurring mineral lodestone are well-known examples of materials that exhibit magnetic properties. Not so familiar, however, is the fact that all substances are influenced to one degree or another by the presence of a magnetic field. This chapter provides a brief description of the origin of magnetic fields and discusses the various magnetic field vectors and magnetic parameters; the phenomena of diamagnetism, paramagnetism, ferromagnetism, and ferrimagnetism; some of the different magnetic materials; and the phenomenon of superconductivity.

8.4.1　Diamagnetism, Paramagnetism and Ferromagnetism

Diamagnetism is a very weak form of magnetism that is nonpermanent and persists only while an external field is being applied. It is induced by a change in the orbital motion of electrons due to an applied magnetic field. The magnitude of the induced magnetic moment is extremely small, and in a direction opposite to that of the applied field. Thus, the relative permeability μ_r is less than unity (however, only very slightly), and the magnetic susceptibility χ_m is negative; that is, the magnitude of the B field within a diamagnetic solid is less than that in a vacuum. The volume susceptibility for diamagnetic solid materials is on the order of -10^{-5}. When placed between the poles of a strong electromagnet, diamagnetic materials are attracted toward regions where the field is weak.

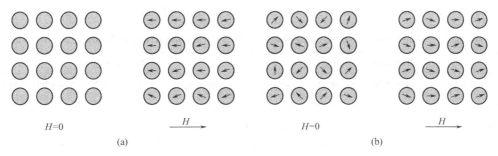

Figure 8.6 (a) The atomic dipole configuration for a diamagnetic material with and without a magnetic field. In the absence of an external field, no dipoles exist; in the presence of a field, dipoles are induced that are aligned opposite to the field direction. (b) Atomic dipole configuration with and without an external magnetic field for a paramagnetic material

Figure 8.6(a) illustrates schematically the atomic magnetic dipole configurations for a diamagnetic material with and without an external field; here, the arrows represent atomic dipole moments, whereas for the preceding discussion, arrows denoted only electron moments. The dependence of B on the external field H for a material that exhibits diamagnetic behavior is presented in Figure 8.7. Table 8.2 gives the susceptibilities of several diamagnetic materials. Diamagnetism is found in all materials; but because it is so weak, it can be observed only when other types of magnetism are totally absent. This form of magnetism is of no practical importance.

For some solid materials, each atom possesses a permanent dipole moment by virtue of incomplete cancellation of electron spin and/or orbital magnetic moments. In the absence of an external magnetic field, the orientations of these atomic magnetic moments are random, such that a piece of material possesses no net macroscopic magnetization. These atomic dipoles are free to rotate, and **paramagnetism** results when they preferentially align, by rotation, with an external field as shown in Figure 8.6(b). These magnetic dipoles are acted on individually with no mutual interaction between adjacent dipoles. Inasmuch as the dipoles align with the external field, they enhance it, giving rise to a relative permeability

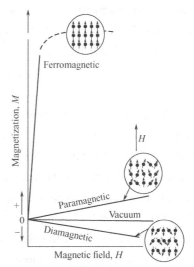

Figure 8.7 Schematic representation of the magnetization M versus the magnetic field strength H for ferromagnetic, paramagnetic and diamagnetic materials. The magnetic susceptibility is positive in paramagnets and ferromagnets, and negative in diamagnets

· 175 ·

μ_r that is greater than unity, and to a relatively small but positive magnetic susceptibility. Susceptibilities for paramagnetic materials range from about 10^{-5} to 10^{-2} (Table 8.2). A schematic B-versus-H curve for a paramagnetic material is also shown in Figure 8.7.

Both diamagnetic and paramagnetic materials are considered to be nonmagnetic because they exhibit magnetization only when in the presence of an external field. Also, for both, the flux density B within them is almost the same as it would be in a vacuum.

Table 8.2 Room-Temperature Magnetic Susceptibilities for Diamagnetic and Paramagnetic Materials

Diamagnetics		Paramagnetics	
Material	Susceptibility χ_m (volume) (SI units)	Material	Susceptibility χ_m (volume) (SI units)
Aluminum oxide	-1.81×10^{-5}	Aluminum	2.07×10^{-5}
Copper	-0.96×10^{-5}	Chromium	3.13×10^{-4}
Gold	-3.44×10^{-5}	Chromium chloride	1.51×10^{-3}
Mercury	-2.85×10^{-5}	Manganese sulfate	3.70×10^{-3}
Silicon	-0.41×10^{-5}	Molybdenum	1.19×10^{-4}
Silver	-2.38×10^{-5}	Sodium	8.48×10^{-6}
Sodium chloride	-1.41×10^{-5}	Titanium	1.81×10^{-4}
Zinc	-1.56×10^{-5}	Zirconium	1.09×10^{-4}

Certain metallic materials possess a permanent magnetic moment in the absence of an external field, and manifest very large and permanent magnetizations. These are the characteristics of **ferromagnetism**, and they are displayed by the transition metals iron (as BCC ferrite), cobalt, nickel, and some of the rare earth metals such as gadolinium (Gd). Magnetic susceptibilities as high as 10^6 are possible for ferromagnetic materials. Consequently, $H \ll M$, and from Equation 8.40 we write

$$B \cong \mu_0 M \qquad (8.6)$$

Permanent magnetic moments in ferromagnetic materials result from atomic magnetic moments due to electron spin-uncancelled electron spins as a consequence of the electron structure. There is also an orbital magnetic moment contribution that is small in comparison to the spin moment. Furthermore, in a ferromagnetic material, coupling interactions cause net spin magnetic moments of adjacent atoms to align with one another, even in the absence of an external field. The origin of these coupling forces is not completely understood, but it is thought to arise from the electronic structure of the metal. This mutual spin alignment exists over relatively large volume regions of the crystal called **domains.**

The maximum possible magnetization, or **saturation magnetization M_s** of a ferromagnetic material represents the magnetization that results when all the magnetic dipoles in a solid piece are mutually aligned with the external field; there is also a corresponding saturation flux density B_s. The saturation magnetization is equal to the product of the net magnetic moment for each atom and the number of atoms present. For each of iron, cobalt, and nickel, the net magnetic moments per atom are 2.22, 1.72, and 0.60 Bohr magnetons, respectively.

8.4.2 Antiferromagnetism and Ferrimagnetism

Antiferromagnetism

This phenomenon of magnetic moment coupling between adjacent atoms or ions occurs in materials other than those that are ferromagnetic. In one such group, this coupling results in an antiparallel alignment; the alignment of the spin moments of neighboring atoms or ions in exactly opposite directions is termed **antiferromagnetism.** Manganese oxide (MnO) is one material that displays this behavior. Manganese oxide is a ceramic material that is ionic in character, having both Mn^{2+} and O^{2-} ions. No net magnetic moment is associated with the O^{2-} ions, since there is a total cancellation of both spin and orbital moments. However, the Mn^{2+} ions possess a net magnetic moment that is predominantly of spin origin. These Mn^{2+} ions are arrayed in the crystal structure such that the moments of adjacent ions are antiparallel. This arrangement is represented schematically in Figure

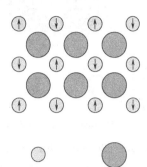

Figure 8.8 Schematic representation of antiparallel alignment of spin magnetic moments for antiferromagnetic manganese oxide

8.8. Obviously, the opposing magnetic moments cancel one another, and, as a consequence, the solid as a whole possesses no net magnetic moment.

Ferrimagnetism

Some ceramics also exhibit a permanent magnetization, termed **ferrimagnetism.** The macroscopic magnetic characteristics of ferromagnets and ferrimagnets are similar; the distinction lies in the source of the net magnetic moments. The principles of ferrimagnetism are illustrated with the cubic ferrites. These ionic materials may be represented by the chemical formula MFe_2O_4, in which M represents any one of several metallic elements. The prototype ferrite is Fe_3O_4, the mineral magnetite, sometimes called lodestone.

The formula for Fe_3O_4 may be written as $Fe^{2+}O^{2-}(Fe^{3+})_2(O^{2-})_3$ in which the Fe ions exist in both +2 and +3 valence states in the ratio of 1:2. A net spin magnetic moment exists for each Fe^{2+} and Fe^{3+} ion, which corresponds to 4 and 5 Bohr magnetons, respectively, for the two ion types. Furthermore, the O^{2-} ions are magnetically neutral. There are antiparallel spin-coupling interactions between the Fe ions, similar in character to antiferromagnetism. However, the net ferromagnetic moment arises from the incomplete cancellation of spin moments.

Cubic ferrites have the inverse spinel crystal structure, which is cubic in symmetry, and similar to the spinel structure. The inverse spinel crystal structure might be thought of as having been generated by the stacking of closepacked planes of O^{2-} ions. For one, the coordination number is 4 (tetrahedral coordination); that is, each Fe ion is surrounded by four oxygen nearest neighbors. For the other, the coordination number is 6 (octahedral coordination). With this inverse spinel structure, half the trivalent (Fe^{3+}) ions are situated in octahedral positions, the other half, in tetrahedral positions. The divalent Fe^{2+} ions are

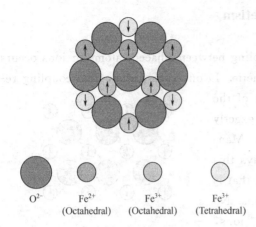

Figure 8.9 Schematic diagram showing the spin magnetic moment configuration

all located in octahedral positions. The critical factor is the arrangement of the spin moments of the Fe ions, as represented in Figure 8.9 and Table 8.3. The spin moments of all the Fe^{3+} ions in the octahedral positions are aligned parallel to one another; however, they are directed oppositely to the Fe^{3+} ions disposed in the tetrahedral positions, which are also aligned. This results from the antiparallel coupling of adjacent iron ions. Thus, the spin moments of all Fe^{3+} ions cancel one another and make no net contribution to the magnetization of the solid. All the Fe^{2+} ions have their moments aligned in the same direction; this total moment is responsible for the net magnetization (see Table 8.4). Thus, the saturation magnetization of a ferrimagnetic solid may be computed from the product of the net spin magnetic moment for each Fe^{2+} ion and the number of Fe^{2+} ions; this would correspond to the mutual alignment of all the Fe^{2+} ion magnetic moments in the Fe_3O_4 specimen.

Table 8.3 The Distribution of Spin Magnetic Moments for Fe^{2+} and Fe^{3+} Ions in a Unit Cell Fe_3O_4 [①]

Cation	Ocathedral Lattice Site	Tetrahedral Lattice Site	Net Magnetic Moment
Fe^{3+}	↑ ↑ ↑ ↑ ↑ ↑ ↑ ↑	↓ ↓ ↓ ↓ ↓ ↓ ↓ ↓	Complete cancellation
Fe^{2+}	↑ ↑ ↑ ↑ ↑ ↑ ↑ ↑	—	↑ ↑ ↑ ↑ ↑ ↑ ↑ ↑

① Each arrow represents the magnetic moment orientation for one of the cations.

Table 8.4 Net Magnetic Moments for Six Cations

Cation	Net Spin Magnetic Moment (Bohr magnetons)
Fe^{3+}	5
Fe^{2+}	4
Mn^{2+}	5
Co^{2+}	3
Ni^{2+}	2
Cu^{2+}	1

Cubic ferrites having other compositions may be produced by adding metallic ions that substitute for some of the iron in the crystal structure. Again, from the ferrite chemical formula, $M^{2+}O^{2-}(Fe^{3+})_2(O^{2-})_3$, in addition to Fe^{2+}, M^{2+} may represent divalent ions such as Ni^{2+}, M_n^{2+}, Co^{2+} and Cu^{2+}, each of which possesses a net spin magnetic moment different from 4; several are listed in Table 8.4. Thus, by adjustment of composition, ferrite compounds having a range of magnetic properties may be produced. For example, nickel ferrite has the formula $NiFe_2O_4$. Other compounds may also be produced containing mixtures of two divalent metal ions such as $(Mn, Mg)Fe_2O_4$, in which the $Mn^{2+} : Mg^{2+}$ ratio may be varied; these are called mixed ferrites.

Ceramic materials other than the cubic ferrites are also ferrimagnetic; these include the hexagonal ferrites and garnets. Hexagonal ferrites have a crystal structure similar to the inverse spinel, with hexagonal symmetry rather than cubic. The chemical formula for these materials may be represented by $AB_{12}O_{19}$, in which A is a divalent metal such as barium, lead, or strontium, and B is a trivalent metal such as aluminum, gallium, chromium, or iron. The two most common examples of the hexagonal ferrites are $PbFe_{12}O_{19}$ and $BaFe_{12}O_{19}$.

The garnets have a very complicated crystal structure, which may be represented by the general formula $M_3Fe_5O_{12}$; here, M represents a rare earth ion such as samarium, europium, gadolinium, or yttrium. Yttrium iron garnet ($Y_3Fe_5O_{12}$), sometimes denoted YIG, is the most common material of this type.

The saturation magnetizations for ferrimagnetic materials are not as high as for ferromagnets. On the other hand, ferrites, being ceramic materials, are good electrical insulators. For some magnetic applications, such as high-frequency transformers, a low electrical conductivity is most desirable.

8.4.3 The Influence of Temperature on Magnetic Behavior

Temperature can also influence the magnetic characteristics of materials. Recall that raising the temperature of solid results in an increase in the magnitude of the thermal vibrations of atoms. The atomic magnetic moments are free to rotate; hence, with rising temperature, the increased thermal motion of the atoms tends to randomize the directions of any moments that may be aligned.

For ferromagnetic, antiferromagnetic, and ferrimagnetic materials, the atomic thermal motions counteract the coupling forces between the adjacent atomic dipole moments, causing some dipole misalignment, regardless of whether an external field is present. This results in a decrease in the saturation magnetization for both ferro and ferrimagnets. The saturation magnetization is a maximum at 0K, at which temperature the thermal vibrations are a minimum. With increasing temperature, the saturation magnetization diminishes gradually and then abruptly drops to zero at what is called the **Curie temperature T_c.**

Antiferromagnetism is also affected by temperature; this behavior vanishes at what is called the *Néel temperature*. At temperatures above this point, antiferromagnetic materials also become paramagnetic.

8.4.4 Domains, Hysteresis and Magnetic Anisotropy

Any ferromagnetic or ferrimagnetic material that is at a temperature below is composed of small-volume regions in which there is a mutual alignment in the same direction of all magnetic dipole moments. Such a region is called a domain, and each one is magnetized to its saturation magnetization. Adjacent domains are separated by domain boundaries or walls, across which the direction of magnetization gradually changes. Normally, domains are microscopic in size, and for a polycrystalline specimen, each grain may consist of more than a single domain. Thus, in a macroscopic piece of material, there will be a large number of domains, and all may have different magnetization orientations. The magnitude of the M field for the entire solid is the vector sum of the magnetizations of all the domains, each domain contribution being weighted by its volume fraction. For an unmagnetized specimen, the appropriately weighted vector sum of the magnetizations of all the domains is zero.

Flux density B and field intensity H are not proportional for ferromagnets and ferrimagnets. If the material is initially unmagnetized, then B varies as a function of H as shown in Figure 8.10. The curve begins at the origin, and as H is increased, the B field begins to increase slowly, then more rapidly, finally leveling off and becoming independent of H. This maximum value of B is the saturation flux density B_s, and the corresponding magnetization is the saturation magnetization M_s, mentioned previously. Since the permeability μ is the slope of the B-versus-H curve, note from Figure 8.10 that the permeability changes with and is dependent on H. On occasion, the slope of the B-versus-H curve at $H=0$ is specified as a material property, which is termed the *initial permeability* μ_i, as indicated in Figure 8.10.

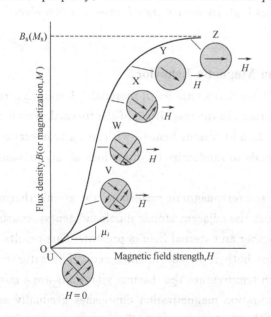

Figure 8.10 The B-versus-H behavior for a ferromagnetic or ferrimagnetic material that was initially unmagnetized. Domain configurations during several stages of magnetization are represented. Saturation flux density B_s, magnetization M_s, and initial permeability μ_i

As an H field is applied, the domains change shape and size by the movement of domain boundaries. Schematic domain structures are represented in the insets (labeled U through Z) at several points along the B-versus-H curve in Figure 8.10. Initially, the moments of the constituent domains are randomly oriented such that there is no net B (or M) field (inset U). As the external field is applied, the domains that are oriented in directions favorable to (or nearly aligned with) the applied field grow at the expense of those that are unfavorably oriented (insets V through X). This process continues with increasing field strength until the macroscopic specimen becomes a single domain, which is nearly aligned with the field (inset Y). Saturation is achieved when this domain, by means of rotation, becomes oriented with the H field (inset Z). Alteration of the domain structure with magnetic field for an iron single crystal is shown in the chapter-opening photographs for this chapter.

From saturation, point S in Figure 8.11, as the H field is reduced by reversal of field direction, the curve does not retrace its original path. A **hysteresis** effect is produced in which the B field lags behind the applied H field, or decreases at a lower rate. At zero H field (point R on the curve), there exists a residual B field that is called the **remanence**, or remanent flux density, B_r; the material remains magnetized in the absence of an external H field.

Hysteresis behavior and permanent magnetization may be explained by the motion of domain walls. Upon reversal of the field direction from saturation (point S in Figure 8.11), the process by which the domain structure changes is reversed. First, there is a rotation of the single domain with the reversed field. Next, domains having

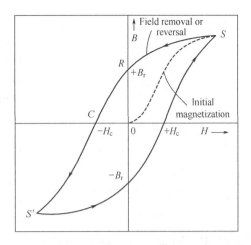

Figure 8.11 Magnetic flux density versus the magnetic field strength for a ferromagnetic material that is subjected to forward and reverse saturations (points S and S'). The hysteresis loop is represented by the solid curve; the dashed curve indicates the initial magnetization. The remanence B_r and the coercive force H_c are also shown

magnetic moments aligned with the new field form and grow at the expense of the former domains. Critical to this explanation is the resistance to movement of domain walls that occurs in response to the increase of the magnetic field in the opposite direction; this accounts for the lag of B with H, or the hysteresis. When the applied field reaches zero, there is still some net volume fraction of domains oriented in the former direction, which explains the existence of the remanence B_r.

To reduce the B field within the specimen to zero (point C on Figure 8.11), an H field of magnitude $-H_c$ must be applied in a direction opposite to that of the original field; H_c is called the **coercivity**, or sometimes the coercive force. Upon continuation of the applied field in this reverse direction, as indicated in the figure, saturation is ultimately achieved in the opposite sense, corresponding to point S'. A second reversal of the field to the point of the initial saturation (point S) completes the symmetrical hysteresis loop and also yields both a negative remanence $(-B_r)$ and a positive coercivity $(+H_c)$.

The magnetic hysteresis curves discussed in the previous section will have different shapes depending on various factors: ①whether the specimen is a single crystal or polycrystalline; ②if polycrystalline, any preferred orientation of the grains; ③the presence of pores or second-phase particles; and④other factors such as temperature and, if a mechanical stress is applied, the stress state.

8.4.5 Superconductivity

The resistivity in superconductors becomes immeasurably small or virtually zero below a critical temperature, T_c, as shown in Figure 8.12. About 27 elements, numerous alloys,

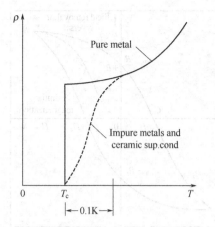

Figure 8.12 Schematic representation of the resistivity of pure and compound superconducting materials. T_c is the critical or transition temperature, below which superconductivity commences

ceramic materials (containing copper oxide), and organic compounds (based, for example, on selenium or sulfur) have been found to possess this property (see Table 8.5). It is estimated that the conductivity of superconductors below T_c is about 10^{20} 1/Ω cm.

The transition temperatures where superconductivity starts range from 0.01K (for tungsten) up to about 125 K (for ceramic superconductors). Of particular interest are materials whose T_c is above 77 K, that is, the boiling point of liquid nitrogen, which is more readily available than other coolants. Among the so-called *high-T_c superconductors* are the 1-2-3 *compounds* such as $YBa_2Cu_3O_{7-x}$ whose molar ratios of rare earth to alkaline earth to copper relate as 1 : 2 : 3. Their transition temperatures range from 40 to 134 K. Ceramic superconductors have an orthorhombic, layered, perovskite crystal structure which contains two-dimensional sheets and periodic oxygen vacancies (The superconductivity exists only parallel to these layers, that is, it is anisotropic). The first superconducting material was found by H. K. Onnes in 1911 in mercury which has a T_c of 4.15 K.

Table 8.5 Critical Temperatures of Some Superconducting Materials

Materials	T_c/K	Remarks
Tungsten	0.01	—
Mercury	4.15	H. K. Onnes(1911)
Sulfur-based organic superconductor	8	S. S. P. Parkin et al. (1983)
Nb_3Sn and Nb-Ti	9	Bell Labs(1961), Type II
V_3Si	17.1	J. K. Hulm(1953)
Nb_3Ge	23.2	(1973)
La-Ba-Cu-O	40	Bednorz and müller(1986)
$YBa_2Cu_3O_{7-x}$	92	Wu, Chu, and others(1987)
$RBa_2Cu_3O_{7-x}$	~92	R=Gd, Dy, Ho, Er, Tm, Yb, Lu
$Bi_2Sr_2Ca_2Cu_3O_{10+\delta}$	113	Maeda et al. (1988)
$Tl_2CaBa_2Cu_2O_{10+\delta}$	125	Hermann et al. (1988)
$HgBa_2Ca_2Cu_3O_{8+\delta}$	134	R. Ott et al. (1995)

A high magnetic field or a high current density may eliminate superconductivity. In *Type I superconductors*, the annihilation of the superconducting state by a magnetic field, that is, the transition between superconducting and normal states, occurs sharply; Figure 8.13(a). The critical field strength H_c, above which superconductivity ceases, is relatively low. The destruction of the superconducting state in *Type II superconductors* occurs instead, more gradually, i.e., in a range between H_{c_1} and H_{c_2}, where H_{c2} is often 100 times larger than H_{c_1} [Figure 8.13(b)]. In the interval between H_{c_1} and H_{c_2}, normal conduc-

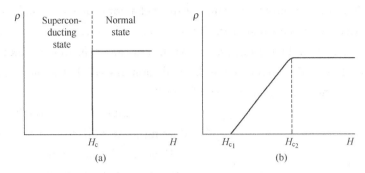

Figure 8.13 Schematic representation of the resistivity of (a) Type I (or soft) and (b) Type II (or hard) superconductors in an external magnetic field. The solids behave like normal conductors above H_c and H_{c2} respectively

ting areas, called *vortices*, and superconducting regions are interspersed. The terms "Type I" and "Type II" superconductors are occasionally also used when a distinction between abrupt and gradual transition with respect to temperature is described; see Figure 8.13. In alloys and ceramic superconductors, a temperature spread of about 0.1 K has been found whereas pure gallium drops its resistance within 10^{-5} K.

Type II superconductors are utilized for strong electromagnets employed, for example, in magnetic resonance imaging devices (used in medicine), high-energy particle accelerators, and electric power storage devices. (An electric current induced into a loop consisting of a superconducting wire continues to flow for an extended period of time without significant decay.) Further potential applications are lossless power transmission lines; highspeed levitation trains; faster, more compact computers; and switching devices, called *cryotrons*, which are based on the destruction of the superconducting state in a strong magnetic field. Despite their considerably higher transition temperatures, ceramic superconductors have not yet revolutionized current technologies, mainly because of their still relatively low T_c, their brittleness, their relatively small capability to carry high current densities, and their environmental instability. These obstacles may be overcome eventually, however, by using other materials, for example, compounds based on bismuth, etc., or by producing thin-film superconductors. At present, most superconducting electromagnets are manufactured by using niobium—titanium alloys which are ductile and thus can be drawn into wires.

Superconductivity is basically an electrical phenomenon; however, its discussion has been deferred to this point because there are magnetic implications relative to the superconducting state, and, in addition, superconducting materials are used primarily in magnets capable of generating high fields.

As most high-purity metals are cooled down to temperatures nearing 0 K, the electrical resistivity decreases gradually, approaching some small yet finite value that is characteristic of the particular metal. There are a few materials, however, for which the resistivity, at a very low temperature, abruptly plunges from a finite value to one that is virtually zero and remains there upon further cooling. Materials that display this latter behavior are called *superconductors*, and the temperature at which they attain **superconductivity** is called the critical temperature T_c. The resistivity—temperature behaviors for superconductive and nonsuperconductive materials are contrasted in Figure 8.14. The critical temperature varies from superconductor to superconductor but lies between less than 1 K and approximately 20 K for metals and metal alloys. Recently, it has been demonstrated that some complex oxide ceramics have critical temperatures in excess of 100 K.

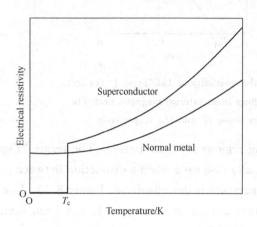

Figure 8.14 Temperature dependence of the electrical resistivity for normally conducting and superconducting materials in the vicinity of 0 K

At temperatures below T_c, the superconducting state will cease upon application of a sufficiently large magnetic field H_c, termed the critical field which depends on temperature and decreases with increasing temperature. The same may be said for current density; that is, a critical applied current density J_c exists below which a material is superconductive.

The superconductivity phenomenon has been satisfactorily explained by means of a rather involved theory. In essence, the superconductive state results from attractive interactions between pairs of conducting electrons; the motions of these paired electrons become coordinated such that scattering by thermal vibrations and impurity atoms is highly inefficient. Thus, the resistivity, being proportional to the incidence of electron scattering, is zero.

On the basis of magnetic response, superconducting materials may be divided into two classifications designated as type I and type II. Type I materials, while in the superconducting state, are completely diamagnetic; that is, all of an applied magnetic field will be excluded from the body of material, a phenomenon known as the *Meissner effect*, which is illustrated in Figure 8.15. As H is increased, the material remains diamagnetic until the critical magnetic field H_c is reached. At this point, conduction becomes normal, and complete magnetic flux penetration takes place. Several metallic elements including aluminum, lead, tin, and mercury belong to the type I group.

Type II superconductors are completely diamagnetic at low applied fields, and field exclusion is total. However, the transition from the superconducting state to the normal state is gradual and occurs between lower critical and upper critical fields, designated H_{c_1} and H_{c_2} respectively. The magnetic flux lines begin to penetrate into the body of material at and with increasing applied magnetic field, this penetration continues; at H_{c_2}, field penetration is complete. For fields between H_{c_1} and H_{c_2}, the material exists in what is termed a mixed state-both normal and superconducting regions are present.

Type II superconductors are preferred over type I for most practical applications by virtue of their higher critical temperatures and critical magnetic fields. At present, the three most commonly utilized superconductors are niobium-zirconium (Nb-Zr) and niobium-titanium (Nb-Ti) alloys and the niobium-tin intermetallic compound Nb_3Sn. Table 8.6 lists sev-

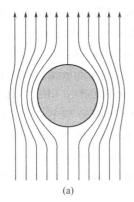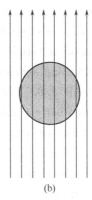

(a)　　　　　　　　　　(b)

Figure 8.15 Representation of the Meissner effect. (a) While in the superconducting state, a body of material (circle) excludes a magnetic field (arrows) from its interior. (b) The magnetic field penetrates the same body of material once it becomes normally conductive

eral type Ⅰ and Ⅱ superconductors, their critical temperatures, and their critical magnetic flux densities.

Table 8.6　Critical Temperatures and Magnetic Fluxes for Selected Superconducting Materials

Material	Critical Temperature T_c (K)	Gritical Magnetic Flux Density B_c (tesla)[1]
Elements[2]		
Tungsten	0.02	0.0001
Titanium	0.40	0.0056
Aluminum	1.18	0.0105
Tin	3.72	0.0035
Mercury(α)	4.15	0.0411
Lead	7.19	0.0803
Compounds and Alloys[2]		
Nb-Ti alloy	10.2	12
Nb-Zr alloy	10.8	11
$PbMo_6S_8$	14.0	45
V_3Ga	16.5	22
Nb_3Sn	18.3	22
Nb_3Al	18.9	32
Nb_3Ge	23.0	30
Ceramic Compounds		
$YBa_2Cu_3O_7$	92	—
$Bi_2Sr_2Ca_2Cu_3O_{10}$	110	—
$Tl_2Ba_2Ca_2Cu_3O_{10}$	125	—
$HgBa_2Ca_2Cu_2O_8$	153	—

[1] The critical magnetic flux density ($\mu_0 H_c$) for the elements was measured at 0 K. For alloys and compounds, the flux is taken as $\mu_0 H_{c2}$ (in teslas), measured at 0 K.

[2] Source: Adapted with permission from *Materials at Low Temperature*, R. P. Reed and A. F. Clark (Editors), American Society for Metals, Metals Park, OH, 1983.

Recently, the family of ceramic materials that are normally electrically insulative have been found to be superconductors with inordinately high critical temperatures. Initial research has centered on yttrium barium copper oxide, $YBa_2Cu_3O_7$, which has a critical temperature of about 92 K. This material has a complex perovskite type crystal structure. New superconducting ceramic materials reported to have even higher critical temperatures have been and are currently being developed. Several of these materials and their critical temperatures are listed in Table 8.6. The technological potential of these materials is extremely promising inasmuch as their critical temperatures are above 77 K, which permits the use of liquid nitrogen, a very inexpensive coolant in comparison to liquid hydrogen and liquid helium. These new ceramic superconductors are not without drawbacks, chief of which is their brittle nature. This characteristic limits the ability of these materials to be fabricated into useful forms such as wires.

Some of the areas being explored include ①electrical power transmission through superconducting materials-power losses would be extremely low, and the equipment would operate at low voltage levels; ②magnets for high-energy particle accelerators; ③higher-speed switching and signal transmission for computers; and ④high-speed magnetically levitated trains, wherein the levitation results from magnetic field repulsion. The chief deterrent to the widespread application of these superconducting materials is, of course, the difficulty in attaining and maintaining extremely low temperatures. Hopefully, this problem will be overcome with the development of the new generation of superconductors with reasonably high critical temperatures.

8.5 Optical Properties of Materials

By "optical property" is meant a material's response to exposure to electromagnetic radiation and, in particular, to visible light. This chapter first discusses some of the basic principles and concepts relating to the nature of electromagnetic radiation and its possible interactions with solid materials. Next to be explored are the optical behaviors of metallic and nonmetallic materials in terms of their absorption, reflection, and transmission characteristics. The final sections outline luminescence, photoconductivity, and light amplification by stimulated emission of radiation (laser), the practical utilization of these phenomena, and optical fibers in communications.

8.5.1 Interaction of Light with Matter

The most apparent properties of metals, their luster and their color, have been known to mankind since materials were known. Because of these properties, metals were already used in antiquity for mirrors and jewelry. The color was utilized 4000 years ago by the ancient Chinese as a guide to determine the composition of the melt of copper alloys; the hue of a preliminary cast indicated whether the melt, from which bells or mirrors were to be made, already had the right tin content.

The German poet Goethe was probably the first one who explicitly spelled out 200 years ago in his *Treatise on Color* that *color* is not an absolute property of matter (such as the resistivity), but requires a living being for its perception and description. Goethe realized that the perceived color of a region in the visual field depends not only on the properties of light coming from that region, but also on the light coming from the rest of the visual field. Applying Goethe's findings, it was possible to explain qualitatively the color of, say, gold in simple terms. Goethe wrote: "If the color *blue* is removed from the spectrum, then blue, violet, and green are missing and red and yellow remain." Thin gold films are bluish-green when viewed in transmission. These colors are missing in reflection. Consequently, gold appears reddish-yellow. On the other hand, Newton stated quite correctly in his "Opticks" that light rays are not colored. The nature of color remained, however, unclear.

This chapter treats the optical properties from a completely different point of view. Measurable quantities such as the index of refraction or the reflectivity and their spectral variations are used to characterize materials. In doing so, the term "color" will almost completely disappear from our vocabulary. Instead, it will be postulated that the interactions of light with the electrons of a material are responsible for the optical properties.

At the beginning of the 20th century, the study of the interactions of light with matter (black-body radiation, etc.) laid the foundations for quantum theory. Today, optical methods are among the most important tools for elucidating the electron structure of matter. Most recently, a number of optical devices such as lasers, photodetectors, waveguides, etc., have gained considerable technological importance. They are used in telecommunication, fiber optics, CD players, laser printers, medical diagnostics, night viewing, solar applications, optical computing, and for optoelectronic purposes. Traditional utilizations of optical materials for windows, antireflection coatings, lenses, mirrors, etc., should be likewise mentioned.

We perceive light intuitively as a wave (specifically, an electromagnetic wave) that travels in undulations from a given source to a point of observation. The color of the light is related to its wavelength. Many crucial experiments, such as diffraction, interference, and dispersion, clearly confirm the wavelike nature of light. Nevertheless, at least since the discovery of the photoelectric effect in 1887 by Hertz, and its interpretation in 1905 by Einstein, do we know that light also has a particle nature (The photoelectric effect describes the emission of electrons from a metallic surface after it has been illuminated by light of appropriately high energy, e.g., by blue light). Interestingly enough, Newton, about 300 years ago, was a strong proponent of the particle concept of light. His original ideas, however, were in need of some refinement, which was eventually provided in 1901 by quantum theory.

Light comprises only an extremely small segment of the entire electromagnetic spectrum, which ranges from radio waves via microwaves, infrared, visible, ultraviolet, X-rays, to γ rays, as depicted in Figure 8.16. Many of the considerations which will be advanced in this chapter are therefore also valid for other wavelength ranges, i.e., for radio waves or X-rays.

8.5.2 Atomic and Electronic Interactions

The optical phenomena that occur within solid materials involve interactions between the

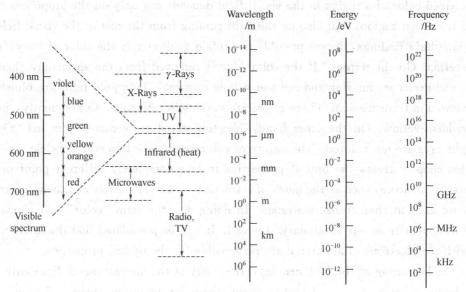

Figure 8.16 The spectrum of electromagnetic radiation. Note the small segment of this spectrum that is visible to human eyes

electromagnetic radiation and atoms, ions, and/or electrons. Two of the most important of these interactions are electronic polarization and electron energy transitions.

Electronic Polarization

One component of an electromagnetic wave is simply a rapidly fluctuating electric field. For the visible range of frequencies, this electric field interacts with the electron cloud surrounding each atom within its path in such a way as to induce electronic polarization, or to shift the electron cloud relative to the nucleus of the atom with each change in direction of electric field component. Two consequences of this polarization are: ① some of the radiation energy may be absorbed, and ② light waves are retarded in velocity as they pass through the medium.

Electron Transitions

The absorption and emission of electromagnetic radiation may involve electron transitions from one energy state to another. For the sake of this discussion, consider an isolated atom, the electron energy diagram for which is represented in Figure 8.17. An electron may be excited from an occupied state at energy E_2 to a vacant and higher-lying one, denoted E_4, by the absorption of a photon of energy. The change in energy experienced by the electron, ΔE depends on the radiation frequency as follows:

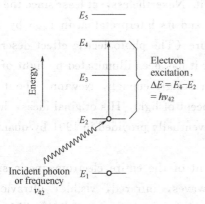

Figure 8.17 For an isolated atom, a schematic illustration of photon absorption by the excitation of an electron from one energy state to another. The energy of the photon must be exactly equal to the difference in energy between the two states $(E_4 - E_2)$.

$$\Delta E = h\nu \qquad (8.7)$$

where, again, h is Planck's constant. At this point it is important that several concepts be understood. First, since the energy states for the atom are discrete, only specific ΔE's exist between the energy levels; thus, only photons of frequencies corresponding to the possible ΔE's for the atom can be absorbed by electron transitions. Furthermore, all of a photon's energy is absorbed in each excitation event.

A second important concept is that a stimulated electron cannot remain in an **excited state** indefinitely; after a short time, it falls or decays back into its **ground state**, or unexcited level, with a reemission of electromagnetic radiation. Several decay paths are possible, and these are discussed later. In any case, there must be a conservation of energy for absorption and emission electron transitions.

As the ensuing discussions show, the optical characteristics of solid materials that relate to absorption and emission of electromagnetic radiation are explained in terms of the electron band structure of the material and the principles relating to electron transitions, as outlined in the preceding two paragraphs.

Optical Properties of Metals

Consider the electron energy band schemes for metals as illustrated in Figure 18.5; in both cases a high-energy band is only partially filled with electrons. Metals are opaque because the incident radiation having frequencies within the visible range excites electrons into unoccupied energy states above the Fermi energy, as demonstrated in Figure 8.18(a); as a consequence, the incident radiation is absorbed, in accordance with Equation (8.8). Total absorption is within a very thin outer layer, usually less than 0.1μm; thus only metallic films thinner than 0.1μm are capable of transmitting visible light. All frequencies of visible light are absorbed by metals because of the continuously available empty electron states, which permit electron transitions as in Figure 8.18(a).

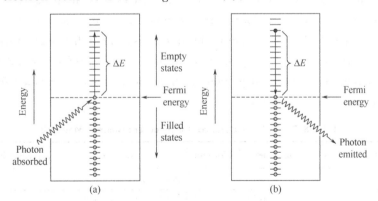

Figure 8.18 (a) Schematic representation of the mechanism of photon absorption for metallic materials in which an electron is excited into a higher-energy unoccupied state. The change in energy of the electron ΔE is equal to the energy of the photon. (b) Reemission of a photon of light by the direct transition of an electron from a high to a low energy state

In fact, metals are opaque to all electromagnetic radiation on the low end of the frequency spectrum, from radio waves, through infrared, the visible, and into about the middle of the ultraviolet radiation. Metals are transparent to high-frequency (x- and -ray) radiation.

Most of the absorbed radiation is reemitted from the surface in the form of visible light of the same wavelength, which appears as reflected light; an electron transition accompanying reradiation is shown in Figure 8.18(b). The reflectivity for most metals is between 0.90 and 0.95; some small fraction of the energy from electron decay processes is dissipated as heat.

Since metals are opaque and highly reflective, the perceived color is determined by the wavelength distribution of the radiation that is reflected and not absorbed. A bright silvery appearance when exposed to white light indicates that the metal is highly reflective over the entire range of the visible spectrum. In other words, for the reflected beam, the composition of these reemitted photons, in terms of frequency and number, is approximately the same as for the incident beam. Aluminum and silver are two metals that exhibit this reflective behavior. Copper and gold appear red-orange and yellow, respectively, because some of the energy associated with light photons having short wavelengths is not reemitted as visible light.

Optical Properties of Nonmetals

By virtue of their electron energy band structures, nonmetallic materials may be transparent to visible light. Therefore, in addition to reflection and absorption, refraction and transmission phenomena also need to be considered.

8.5.3 Refraction, Reflection, Absorption and Transmission

Light that is transmitted into the interior of transparent materials experiences a decrease in velocity, and, as a result, is bent at the interface; this phenomenon is termed **refraction**.

For crystalline ceramics that have cubic crystal structures, and for glasses, the index of refraction is independent of crystallographic direction (i.e., it is isotropic). Noncubic crystals, on the other hand, have an anisotropic n; that is, the index is greatest along the directions that have the highest density of ions. Table 8.7 gives refractive indices for several glasses, transparent ceramics, and polymers. Average values are provided for the crystalline ceramics in which n is anisotropic.

Table 8.7 Refractive Indices for Some Transparent Materials

Material	Average Index of Refraction	Material	Average Index of Refraction
Ceramics		**Polymers**	
Silica glass	1.458	Polytetrafluoroethylene	1.35
Borosilicate(Pyrex)glass	1.47	Polymethyl methacrylate	1.49
Soda-lime glass	1.51	Polypropylene	1.49
Quartz(SiO_2)	1.55	Polyethylene	1.51
Dense optical flint glass	1.65	Polystyrene	1.60
Spinel($MgAl_2O_4$)	1.72		
Periclase(MgO)	1.74		
Corundum(Al_2O_3)	1.76		

When light radiation passes from one medium into another having a different index of refraction, some of the light is scattered at the interface between the two media even if both are transparent. The **reflectivity** R represents the fraction of the incident light that is reflected at the interface, or

$$R = \frac{I_R}{I_0} \tag{8.8}$$

where I_0 and I_R are the intensities of the incident and reflected beams, respectively.

Nonmetallic materials may be opaque or transparent to visible light; and, if transparent, they often appear colored. In principle, light radiation is absorbed in this group of materials by two basic mechanisms, which also influence the transmission characteristics of these nonmetals. One of these is electronic polarization. **Absorption** by electronic polarization is important only at light frequencies in the vicinity of the relaxation frequency of the constituent atoms. The other mechanism involves valence band-conduction band electron transitions, which depend on the electron energy band structure of the material.

Absorption of a photon of light may occur by the promotion or excitation of an electron from the nearly filled valence band, across the band gap, and into an empty state within the conduction band, a free electron in the conduction band and a hole in the valence band are created.

The phenomena of absorption, reflection, and transmission may be applied to the passage of light through a transparent solid, as shown in Figure 8.19. For an incident beam of intensity I_0 that impinges on the front surface of a specimen of thickness l and absorption coefficient β, the **transmitted intensity** at the back face I_T is

$$I_T = I_0 (1-R)^2 e^{-\beta l} \tag{8.9}$$

where R is the reflectance; for this expression, it is assumed that the same medium exists outside both front and back faces. The derivation of Equation (8.9) is left as a homework problem.

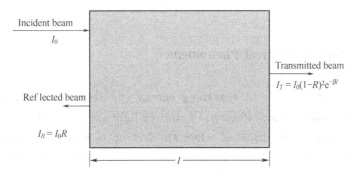

Figure 8.19 The transmission of light through a transparent medium for which there is reflection at front and back faces, as well as absorption within the medium

Thus, the fraction of incident light that is transmitted through a transparent material depends on the losses that are incurred by absorption and reflection. Again, the sum of the reflectivity R, absorptivity A, and transmissivity T, is unity according to Equation 8.5. Also, each of the variables R, A, and T depends on light wavelength. For example, for light having a wavelength of 0.4m, the fractions transmitted, absorbed, and reflected are approximately 0.90, 0.05, and 0.05, respectively. However, at 0.55m, the respective fractions have shifted to about 0.50, 0.48, and 0.02.

8.5.4 Opacity and Translucency in Insulators

The extent of translucency and opacity for inherently transparent dielectric materials depends to a great degree on their internal reflectance and transmittance characteristics. Many dielectric materials that are intrinsically transparent may be made translucent or even opaque because of interior reflection and refraction. A transmitted light beam is deflected in direction and appears diffuse as a result of multiple scattering events. Opacity results when the scattering is so extensive that virtually none of the incident beam is transmitted, undeflected, to the back surface.

This internal scattering may result from several different sources. Polycrystalline specimens in which the index of refraction is anisotropic normally appear translucent. Both reflection and refraction occur at grain boundaries, which cause a diversion in the incident beam. This results from a slight difference in index of refraction n between adjacent grains that do not have the same crystallographic orientation.

Scattering of light also occurs in two-phase materials in which one phase is finely dispersed within the other. Again, the beam dispersion occurs across phase boundaries when there is a difference in the refractive index for the two phases; the greater this difference, the more efficient is the scattering. Glass-ceramics, which may consist of both crystalline and residual glass phases, will appear highly transparent if the sizes of the crystallites are smaller than the wavelength of visible light, and when the indices of refraction of the two phases are nearly identical (which is possible by adjustment of composition).

As a consequence of fabrication or processing, many ceramic pieces contain some residual porosity in the form of finely dispersed pores. These pores also effectively scatter light radiation.

8.5.5 Applications of Optical Phenomena
Luminescence

Some materials are capable of absorbing energy and then reemitting visible light in a phenomenon called **luminescence**. Photons of emitted light are generated from electron transitions in the solid. Energy is absorbed when an electron is promoted to an excited energy state; visible light is emitted when it falls back to a lower energy state if $1.8 \text{ eV} < h\nu < 3.1 \text{ eV}$. The absorbed energy may be supplied as higher energy electromagnetic radiation causing valence band—conduction band transitions, such as ultraviolet light, or other sources such as high energy electrons, or by heat, mechanical, or chemical energy. Furthermore, luminescence is classified according to the magnitude of the delay time between absorption and reemission events. If reemission occurs for times much less than one second, the phenomenon is termed **fluorescence**; for longer times, it is called **phosphorescence**. A number of materials can be made to fluoresce or phosphoresce; these include some sulfides, oxides, tungstates, and a few organic materials. Ordinarily, pure materials do not display these phenomena, and to induce them, impurities in controlled concentrations must be added.

Luminescence has a number of commercial applications. Fluorescent lamps consist of glass housing, coated on the inside with specially prepared tungstates or silicates. Ultraviolet light is generated within the tube from a mercury glow discharge, which causes the coating to fluoresce and emit white light. The picture viewed on a television screen (cathode ray tube screen) is the product of luminescence. The inside of the screen is coated with a material that fluoresces as an electron beam inside the picture tube very rapidly traverses the screen. Detection of X-rays and γ-rays is also possible; certain phosphors emit visible light or glow when introduced into a beam of the radiation that is otherwise invisible.

Photoconductivity

The conductivity of semiconducting materials depends on the number of free electrons in the conduction band and also the number of holes in the valence band. Thermal energy associated with lattice vibrations can promote electron excitations in which free electrons and/or holes are created. Additional charge carriers may be generated as a consequence of photon-induced electron transitions in which light is absorbed; the attendant increase in conductivity is called **photoconductivity.**

Thus, when a specimen of a photoconductive material is illuminated, the conductivity increases. This phenomenon is utilized in photographic light meters. A photoinduced current is measured, and its magnitude is a direct function of the intensity of the incident light radiation, or the rate at which the photons of light strike the photoconductive material. Of course, visible light radiation must induce electronic transitions in the photoconductive material; cadmium sulfide is commonly utilized in light meters.

Sunlight may be directly converted into electrical energy in solar cells, which also employ semiconductors. The operation of these devices is, in a sense, the reverse of that for the light-emitting diode. A p-n junction is used in which photoexcited electrons and holes are drawn away from the junction, in opposite directions, and become part of an external current.

Optical phenomena are also applied in stimulated electron transitions such as lasers, Coherent and high-intensity light beams and transmission of information such as fiber-optic technology in communications.

Summary

Electrical Properties

The ease with which a material is capable of transmitting an electric current is expressed in terms of electrical conductivity or its reciprocal, resistivity. On the basis of its conductivity, a solid material may be classified as a metal, a semiconductor, or an insulator.

For most materials, an electric current results from the motion of free electrons, which are accelerated in response to an applied electric field. The number of these free electrons depends on the electron energy band structure of the material. An electron band is just a series of electron states that are closely spaced with respect to energy, and one such band may exist for each electron subshell found in the isolated atom. By "electron energy band structure" is meant the manner in which the outermost bands are arranged relative to one another and then filled with electrons. A distinctive band structure type exists for metals, for semiconductors, and for insulators. An electron becomes free by being excited from a filled state in one band, to an available empty state above the Fermi energy. Relatively small energies are required for electron excitations in metals, giving rise to large numbers of free electrons. Larger energies are required for electron excitations in semiconductors and insulators, which accounts for their lower free electron concentrations and smaller conductivity values.

For metallic materials, electrical resistivity increases with temperature, impurity content, and plastic deformation. The contribution of each to the total resistivity is additive.

Semiconductors may be either elements (Si and Ge) or covalently bonded compounds. With these materials, in addition to free electrons, holes (missing electrons in the valence band) may also participate in the conduction process. On the basis of electrical behavior, semiconductors are classified as either intrinsic or extrinsic.

Thermal Properties

This section discussed heat absorption, thermal expansion, and thermal conduction—three important thermal phenomena. Heat capacity represents the quantity of heat required to produce a unit rise in temperature for one mole of a substance; on a per-unit mass basis, it is termed specific heat. Most of the energy assimilated by many solid materials is associated with increasing the vibrational energy of the atoms; contributions to the total heat capacity by other energy-absorptive mechanisms (i. e., increased free-electron kinetic energies) are normally insignificant. For many crystalline solids and at temperatures within the vicinity of 0 K, the heat capacity measured at constant volume varies as the cube of the absolute temperature; in excess of the Debye temperature, C_v becomes temperature independent, assuming a value of approximately $3R$.

Solid materials expand when heated and contract when cooled. The fractional change in length is proportional to the temperature change, the constant of proportionality being the coefficient of thermal expansion. Thermal expansion is reflected by an increase in the average interatomic separation, which is a consequence of the asymmetric nature of the potential energy versus interatomic spacing curve trough. The larger the interatomic bonding energy, the lower is the coefficient of thermal expansion.

Thermal stresses, which are introduced in a body as a consequence of temperature changes, may lead to fracture or undesirable plastic deformation. The two prime sources of thermal stresses are restrained thermal expansion (or contraction) and temperature gradients established during heating or cooling. Thermal shock is the fracture of a body resulting from thermal stresses induced by rapid temperature changes. Because ceramic materials are brittle, they are especially susceptible to this type of failure. The thermal shock resistance of many materials is proportional to the fracture strength and thermal conductivity, and inversely proportional to both the modulus of elasticity and the coefficient of thermal expansion.

Magnetic Properties

The macroscopic magnetic properties of a material are a consequence of interactions between an external magnetic field and the magnetic dipole moments of the constituent atoms. Associated with each individual electron are both orbital and spin magnetic moments. The net magnetic moment for an atom is just the sum of the contributions of each of its electrons, wherein there will be spin and orbital moment cancellation of electron pairs.

Diamagnetism results from changes in electron orbital motion that are induced by an external field. The effect is extremely small and in opposition to the applied field. All materials are diamagnetic. Paramagnetic materials are those having permanent atomic dipoles, which are acted on individually and are aligned in the direction of an external field. Since the magnetizations are relatively small and persist only while an applied field is present, diamagnetic and paramagnetic materials are considered to be nonmagnetic.

Large and permanent magnetizations may be established within the ferromagnetic metals (Fe, Co, Ni). Atomic magnetic dipole moments are of spin origin, which are coupled and mutually aligned with moments of adjacent atoms.

Antiparallel coupling of adjacent cation spin moments is found for some ionic materials. Those in which there is total cancellation of spin moments are termed antiferromagnetic. With ferrimagnetism, permanent magnetization is possible because spin moment cancellation is incomplete. For cubic ferrites, the net magnetization results from the divalent ions (e.g., Fe^{2+}) that reside on octahedral lattice sites, the spin moments of which are all mutually aligned.

With rising temperature, increased thermal vibrations tend to counteract the dipole coupling forces in ferromagnetic and ferrimagnetic materials. Consequently, the saturation magnetization gradually diminishes with temperature, up to the Curie temperature, at which point it drops to near zero; above these materials are paramagnetic.

Below its Curie temperature, a ferromagnetic or ferrimagnetic material is composed of domains—small-volume regions wherein all net dipole moments are mutually aligned and the magnetization is saturated. The total magnetization of the solid is just the appropriately weighted vector sum of the magnetizations of all these domains. As an external magnetic field is applied, domains having magnetization vectors oriented in the direction of the field grow at the expense of domains that have unfavorable magnetization orientations. At total saturation, the entire solid is a single domain and the magnetization is aligned with the field direction. The change in domain structure with increase or reversal of a magnetic field is accomplished by the motion of domain walls. Both hysteresis (the lag of the B field behind the applied H field) as well as permanent magnetization (or remanence) result from the resistance to movement of these domain walls.

Superconductivity has been observed in a number of materials, in which, upon cooling and in the vicinity of absolute zero temperature, the electrical resistivity vanishes. The superconducting state ceases to exist if temperature, magnetic field, or current density exceeds the critical value. For type I superconductors, magnetic field exclusion is complete below a critical field, and field penetration is complete once is exceeded. This penetration is gradual with increasing magnetic field for type II materials. New complex oxide ceramics are being developed with relatively high critical temperatures, which allow inexpensive liquid nitrogen to be used as a coolant.

Optical Properties

The optical behavior of a solid material is a function of its interactions with electromagnetic radiation having wavelengths within the visible region of the spectrum. Possible interactive phenomena include refraction, reflection, absorption, and transmission of incident light.

Light radiation experiences refraction in transparent materials; that is, its velocity is retarded and the light beam is bent at the interface. Index of refraction is the ratio of the velocity of light in a vacuum to that in the particular medium. The phenomenon of refraction is a consequence of electronic polarization of the atoms or ions, which is induced by the electric field component of the light wave.

Two other important optical phenomena were discussed: luminescence and photoconductivity. With luminescence, energy is absorbed as a consequence of electron excitations, which is reemitted as visible light. The electrical conductivity of some semiconductors may be enhanced by photoinduced electron transitions, whereby additional free electrons and holes are generated.

Important Terms And Concepts

Acceptor state	Extrinsic semiconductor	Ohm's law
Capacitance	Fermi energy	Permittivity
Conduction band	Ferroelectric	Piezoelectric
Conductivity, electrical	Forward bias	Polarization
Dielectric	Free electron	Polarization, electronic
Dielectric constant	Hall effect	Polarization, ionic
Dielectric displacement	Hole	Polarization, orientation
Dielectric strength	Insulator	Rectifying junction
Diode	Integrated circuit	Relaxation frequency
Dipole, electric	Intrinsic semiconductor	Resistivity, electrical
Donor state	Ionic conduction	Reverse bias
Doping	Junction transistor	Semiconductor
Electrical resistance	expansion	Valence band
Electron energy band	Metal	Phonon
Energy band gap	Mobility	Thermal shock
Heat capacity	Specific heat	Thermal stress
Linear coefficient of thermal	Thermal conductivity	Paramagnetism
Antiferromagnetism	Ferromagnetism	Remanence
Coercivity	Hysteresis	Saturation magnetization
Curie temperature	Magnetic field strength	Soft magnetic material
Diamagnetism	Magnetic flux density	Superconductivity
Domain	Magnetic induction	Magnetization
Ferrimagnetism	Magnetic susceptibility	Reflection
Ferrite(ceramic)	Absorption	Refraction
Light-emitting diode(LED)	Luminescence	Translucent
Electroluminescence	Opaque	Transmission
Excited state	Phosphorescence	Transparent
Fluorescence	Photoconductivity	Laser
Ground state	Photon	
Index of refraction		

References

[1] Azaroff, L. V. and J. J. Brophy, *Electronic Processes in Materials*, McGraw-Hill, New York, 1963. Reprinted by TechBooks, Marietta, OH, 1990.

[2] Bube, R. H., *Electrons in Solids*, 3rd edition, Academic Press, San Diego, 1992.

[3] Chaudhari, P., "Electronic and Magnetic Materials," *Scientific American*, Vol. 255, No. 4, October, 1986.

[4] Hummel, R. E., *Electronic Properties of Materials*, 3rd edition, Springer-Verlag, New York, 2000.

[5] Kingery, W. D., H. K. Bowen, and D. R. Uhlmann, *Introduction to Ceramics*, 2nd edition, Wiley, New York, 1976.

[6] Kittel, C., *Introduction to Solid State Physics*, 8th edition, Wiley, New York, 2005.

[7] Kwok, H. L., *Electronic Materials*, PWS Publishers, Boston, 1997.

[8] Livingston, J., *Electronic Properties of Engineering Materials*, Wiley, New York, 1999.

[9] Pierret, R. F. and K. Harutunian, *Semiconductor Device Fundamentals*, Addison-Wesley Longman, Boston, 1996.

[10] Solymar, L. and D. Walsh, *Electrical Properties of Materials*, 7th edition, Oxford University Press, New York, 2004.

[11] L. Brillouin, *Wave Propagation in Periodic Structures*, Dover, New York (1953)

[12] R. E. Hummel, *Electronic Properties of Materials*, 3rd Edition, Springer-Verlag, New York (2001).

[13] Kingery, W. D., H. K. Bowen, and D. R. Uhlmann, *Introduction to Ceramics*, 2nd edition, Wiley, New York, 1976.

[14] Rose, R. M., L. A. Shepard, and J. Wulff, *The Structure and Properties of Materials*, Vol. IV. *Electronic Properties*, Wiley, New York, 1966.

[15] Ziman, J., "The Thermal Properties of Materials," *Scientific American*, Vol. 217, No. 3, September 1967.

[16] Azaroff, L. V. and J. J. Brophy, *Electronic Processes in Materials*, McGraw-Hill Book Company, New York, 1963. Reprinted by CBLS Publishers, Marietta, OH, 1990.

[17] Bozorth, R. M., *Ferromagnetism*, Wiley-IEEE Press, New York/Piscataway, NJ, 1993.

[18] Brockman, F. G., "Magnetic Ceramics—A Review and Status Report," *American Ceramic Society Bulletin*, Vol. 47, No. 2, February 1968.

[19] Chen, C. W., *Magnetism and Metallurgy of Soft Magnetic Materials*, Dover Publications, New York, 1986.

[20] Jiles, D., *Introduction to Magnetism and Magnetic Materials*, Nelson Thornes, Cheltenham, UK, 1998.

[21] Keffer, F., "The Magnetic Properties of Materials," *Scientific American*, Vol. 217, No. 3, September 1967.

[22] Lee, E. W., *Magnetism, An Introductory Survey*, Dover Publications, New York, 1970.

[23] Morrish, A. H., *The Physical Principles of Magnetism*, Wiley-IEEE Press, New York/Piscataway, NJ, 2001.

[24] Azaroff, L. V., and J. J. Brophy, *Electronic Processes in Materials*, McGraw-Hill, New York, 1963. Reprinted by TechBooks, Marietta, OH.

[25] Javan, A., "The Optical Properties of Materials," *Scientific American*, Vol. 217, No. 3, September 1967.

[26] Kingery, W. D., H. K. Bowen, and D. R. Uhlmann, *Introduction to Ceramics*, 2nd edition, Wiley, New York, 1976.

[27] Ralls, K. M., T. H. Courtney, and J. Wulff, *Introduction to Materials Science and Engineering*, Wiley, New York, 1976.

[28] Rowell, J. M., "Photonic Materials," *Scientific American*, Vol. 255, No. 4, October 1986.

Chapter 9 Biomaterials/Nanomaterials/Smart Materials

Learning Objectives

After careful study of this chapter you should be able to do the following:

1. Define the biomaterials and briefly describe the performance and historical background of biomaterials.
2. Define the nanotechnology and nanometer.
3. Briefly describe examples of current achievements for nanotechnology and nanomaterials
4. Briefly describe shape applications of smart materials.

9.1 Biomaterials

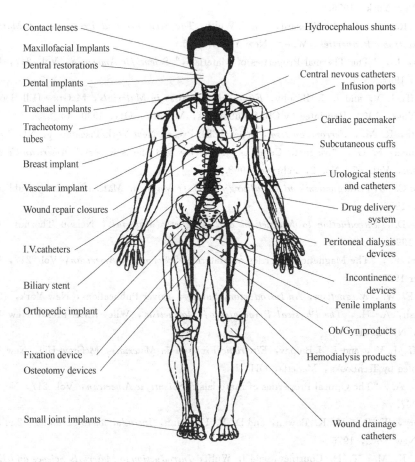

Illustrations of various implants and devices used to replace or enhance the function of diseased or missing tissues and organs.

· 198 ·

9.1.1 Definition of Biomaterials

A *biomaterial* can be defined as any material used to make devices to replace a part or a function of the body in a safe, reliable, economic, and physiologically acceptable manner. Some people refer to materials of biological origin such as wood and bone as biomaterials, but in this book we refer to such materials as "biological materials." A variety of devices and materials is used in the treatment of disease or injury. Commonplace examples include sutures, tooth fillings, needles, catheters, bone plates, etc. A biomaterial is a synthetic material used to replace part of a living system or to function in intimate contact with living tissue. The Clemson University Advisory Board for Biomaterials has formally defined a biomaterial to be "a systemically and pharmacologically inert substance designed for implantation within or incorporation with living systems". These descriptions add to the many ways of looking at the same concept but expressing it in different ways. By contrast, a *biological material* is a material such as bone, skin, or artery produced by a biological system. Artificial materials that simply are in contact with the skin, such as hearing aids and wearable artificial limbs, are not included in our definition of biomaterials since the skin acts as a barrier with the external world.

Because the ultimate goal of using biomaterials is to improve human health by restoring the function of natural living tissues and organs in the body, it is essential to understand relationships among the properties, functions, and structures of biological materials. Thus, three aspects of study on the subject of biomaterials can be envisioned: biological materials, implant materials, and interaction between the two in the body.

The success of a biomaterial or an implant is highly dependent on three major factors: the properties and biocompatibility of the implant (Figure 9.1), the health condition of the recipient, and the competency of the surgeon who implants and monitors its progress. It is

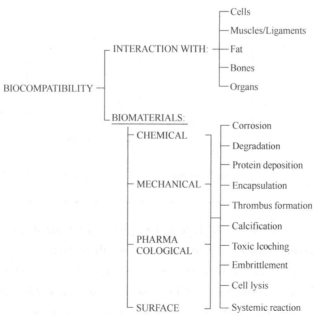

Figure 9.1 Schematic illustration of biocompatibility

easy to understand the requirements for an implant by examining the characteristics that a bone plate must satisfy for stabilizing a fractured femur after an accident. These are:

1. Acceptance of the plate to the tissue surface, i.e., biocompatibility (this is a broad term and includes points 2 and 3)
2. Pharmacological acceptability (nontoxic, nonallergenic, nonimmunogenic, noncarcinogenic, etc.)
3. Chemically inert and stable (no time-dependent degradation)
4. Adequate mechanical strength
5. Adequate fatigue life
6. Sound engineering design
7. Proper weight and density
8. Relatively inexpensive, reproducible, and easy to fabricate and process for large-scale production.

Development of an understanding of the properties of materials that can meet these requirements is one of the goals of this book. The list in Table 9.1 illustrates some of the advantages, disadvantages, and applications of four groups of synthetic (manmade) materials used for implantation. Reconstituted (natural) materials such as collagen have been used for replacements (e.g., arterial wall, heart valve, and skin).

Table 9.1 Class of Materials Used in the Body

Materials	Advantages	Disadvantages	Examples
Polymers(nylon, silicone rubber, polyester, polytetrafuoroethylene, etc)	Resilient Easy to fabricate	Not strong Deforms with time May degrade	Sutures, blood vessels other soft tissues, sutures, hip socket, ear, nose
Metals(Ti and its alloys, Co-Cr alloys, Au, Ag stainless steels, etc.)	Strong, tough ductile	May corrode Dense Difficult to make	Joint replacements, dental root implants, pacer and suture wires, bone plates and screws
Ceramics(alumina zirconia, calcium phosphates including hydroxyapatite, carbon)	Very biocompatible	Brittle Not resilient Weak in tension	Dental and orthopedic implants
Composites(carboncarbon, wire-or fiber-reinforced bone cement)	Strong, tailor-made	Difficult to make	Bone cement, Dental resin

The materials to be used in vivo have to be approved by the FDA (United States Food and Drug Administration). If a proposed material is substantially equivalent to one used before the FDA legislation of 1976, then the FDA may approve its use on a Premarket Approval (PMA) basis. This process, justified by experience with a similar material, reduces the time and expense for the use of the proposed material. Otherwise, the material has to go through a series of "biocompatibility" tests. In general biocompatibility requirements in-

clude:
1. Acute systemic toxicity
2. Cytotoxicity
3. Hemolysis
4. Intravenous toxicity
5. Mutagenicity
6. Oral toxicity
7. Pyrogenicity
8. Sensitization

The guidelines on biocompatibility assessment are given in Table 9.2. The data and documentation requirements for all tests demonstrate the importance of good recordkeeping. It is also important to keep all documents created in the production of materials and devices to be used in vivo within the boundaries of Good Manufacturing Practices (GMP), requiring completely isolated clean rooms for production of implants and devices. The final products are usually sterilized after packaging. The packaged item is normally mass sterilized by γ-radiation or ETO (ethylene oxide gas).

Table 9.2 Guidance on Biocompatibility Assessment

A. Data required to asses suitability

  1. Material characterization. Identify the chemical structure of a material and any potential toxicological hazards. Residue levels. Degradation products. Cumulative effects of each process

  2. Information on prior use. Documented proof of prior use, which would indicate the material(s) suitability

  3. Toxicological data. Results of known biological tests that would aid in assessing potential reaction (adverse or not) during clinical use

B. Supporting documents

  1. Details of application: shape, size, form, plus time in contact and use

  2. Chemical breakdown of all materials involves in the product

  3. A review of all toxicity data on those materials in direct contact with the body tissues

  4. Prior use and details of effects

  5. Toxicity tests[FDA[1] or ISO(International Standard Organization guides)]

  6. Final assessment of all information including toxiological significance

[1] FDA internet address: http://www.fda.gov/cdrh/index.html.
CDRH (Center for Devices and Radiological Health of the FDA) administers medical devices.

Nanotechnology is a rapidly evolving field that involves material structures on a size scale typically 100 nm or less. New areas of biomaterials applications may develop using nanoscale materials or devices. For example, drug delivery methods have made use of a microsphere encapsulation technique. Nanotechnology may help in the design of drugs with more precise dosage, oriented to specific targets or with timed interactions. Nanotechnology may also help to reduce the size of diagnostic sensors and probes. Transplantation of organs can restore some functions that cannot be carried out by artificial materials, or that are better done by a natural organ. For example, in the case of kidney failure many patients can expect to derive benefit from transplantation because an artificial kidney has many disadvantages, including high cost, immobility of the device, maintenance of the dialyzer, and illness due to imperfect filtration. The functions of the liver cannot be assumed by any artificial device or material. Liver transplants have extended the lives of people with liver failure. Organ transplants are widely performed, but their success has been hindered due to social, ethical, and immunological problems.

Since artificial materials are limited in the functions they can perform, and transplants are limited by the availability of organs and problems of immune compatibility, there is current interest in the regeneration or regrowth of diseased or damaged tissue. **Tissue engineering** refers to the growth of a new tissue using living cells guided by the structure of a substrate made of synthetic material. This substrate is called a **scaffold**. The scaffold materials are important since they must be compatible with the cells and guide their growth. Most scaffold materials are biodegradable or resorbable as the cells grow. Most scaffolds are made from natural or synthetic polymers, but for hard tissues such as bone and teeth ceramics such as calcium phosphate compounds can be utilized. The tissue is grown in vitro and implanted in vivo. There have been some clinical successes in repair of injuries to large areas of skin, or small defects in cartilage.

It is imperative that we should know the fundamentals of materials before we can utilize them properly and efficiently. Meanwhile, we also have to know some fundamental properties and functions of tissues and organs. The interactions between tissues and organs with manmade materials have to be more fully elucidated. Fundamentals-based scientific knowledge can be a great help in exploring many avenues of biomaterials research and development.

9.1.2 Performance of Biomaterials

The performance of an implant after insertion can be considered in terms of reliability. For example, there are four major factors contributing to the failure of hip joint replacements. These are fracture, wear, infection, and loosening of implants, as shown in Figure 9.2. If the probability of failure of a given system is assumed to be f, then the reliability, r, can be expressed as

$$r = 1 - f \tag{9.1}$$

Total reliability r_t can be expressed in terms of the reliabilities of each contributing factor for failure:

$$r_t = r_1, r_2, \cdots, r_n \tag{9.2}$$

where $r_1 = 1 - f_1$, $r_2 = 1 - f_2$, and so on.

Equation (9.2) implies that even though an implant has a perfect reliability of one (i.e., $r = 1$), if an infection occurs every time it is implanted then the total reliability of the operation is zero. Actually, the reliability of joint replacement procedures has greatly improved since they were first introduced.

The study of the relationships between the structure and physical properties of biological materials is as important as that of biomaterials, but traditionally this subject has not been treated fully in biologically oriented disciplines. This is due to the fact that in these disci-

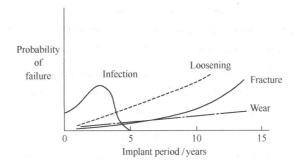

Figure 9.2 A schematic illustration of probability of failure versus implant period for hip joint replacements

plines workers are concerned with the biochemical aspects of function rather than the physical properties of "materials". In many cases one can study biological materials while ignoring the fact that they contain and are made from living cells. For example, in teeth the function is largely mechanical, so that one can focus on the mechanical properties of the natural materials. In other cases the functionality of the tissues or organs is so dynamic that it is meaningless to replace them with biomaterials, e. g., the spinal cord or brain.

9.1.3 Brief Historical Background

Historically speaking, until Dr. J. Lister's aseptic surgical technique was developed in the 1860s, attempts to implant various metal devices such as wires and pins constructed of iron, gold, silver, platinum, etc. were largely unsuccessful due to infection after implantation. The aseptic technique in surgery has greatly reduced the incidence of infection. Many recent developments in implants have centered around repairing long bones and joints. Lane of England designed a fracture plate in the early 1900s using steel, as shown in Figure 9.3(a). Sherman of Pittsburgh modified the Lane plate to reduce the stress concentration by eliminating sharp corners [Figure 9.3(b)]. He used vanadium alloy steel for its toughness and ductility. Subsequently, Stellite

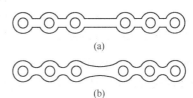

Figure 9.3 Early design of bone fracture plate: (a) Lane, (b) Sherman

□ (Co-Cr-based alloy) was found to be the most inert material for implantation by Zierold in 1924. Soon 18-8 (18 w/o Cr, 8 w/o Ni) and 18-8 sMo (2-4 w/o Mo) stainless steels were introduced for their corrosion resistance, with 18-8sMo being especially resistant to corrosion in saline solution. Later, another alloy (19 w/o Cr, 9 w/o Ni) named Vitallium was introduced into medical practice. A noble metal, tantalum, was introduced in 1939, but its poor mechanical properties and difficulties in processing it from the ore made it unpopular in orthopedics, yet it found wide use in neurological and plastic surgery. During the post-Lister period, the various designs and materials could not be related specifically to the success or failure of an implant, and it became customary to remove any metal implant as soon as possible after its initial function was served.

Fracture repair of the femoral neck was not initiated until 1926, when Hey-Groves used carpenter's screws. Later, Smith-Petersen (1931) designed the first nail with protruding fins to prevent rotation of the femoral head. He used stainless steel but soon changed to Vitallium. Thornton (1937) attached a metal plate to the distal end of the Smith-Petersen nail and secured it with screws for better support. Smith-Petersen later (1939) used an artificial cup over the femoral head in order to create new surfaces to substitute for the diseased joints. He used glass, Pyrex, Bakelite, and Vitallium. The latter was found more biologically compatible, and $30\% \sim 40\%$ of patients gained usable joints. Similar mold arthroplastic surgeries were performed successfully by the Judet brothers of France, who used the first biomechanical designed prosthesis made of an acrylic (methylmethacrylate) polymer (Figure 9.4). The same type of acrylic polymer was also used for corneal replacement in the 1940s and 1950s due to its excellent properties of transparency and biocompatibility.

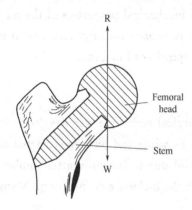

Figure 9.4 The Judet prosthesis for hip surface arthroplasty

Due to the difficulty of surgical techniques and to material problems, cardiovascular implants were not attempted until the 1950s. Blood vessel implants were attempted with rigid tubes made of polyethylene, acrylic polymer, gold, silver, and aluminum, but these soon filled with clot. The major advancement in vascular implants was made by Voorhees, Jaretzta, and Blackmore (1952), when they used a cloth prosthesis made of polyvinyl chloride and polyacrylonitrile and later experimented with nylon, Orlon, Dacron, Teflon, and Ivalon. Through the pores of the various cloths a pseudo- or neointima was formed by tissue ingrowth. This new lining was more compatible with blood than a solid synthetic surface, and it prevented further blood coagulation. Heart valve implantation was made possible only after the development of open-heart surgery in the mid-1950s. Starr and Edwards (1960) made the first commercially available heart valve, consisting of a silicone rubber ball poppet in a metal strut (Figure 9.5). Concomitantly, artificial heart and heart assist devices have been developed. Table 9.3 gives a brief summary of historical developments relating to implants.

Figure 9.5 An early model of the Starr-Edwards heart valve made of a silicone rubber ball and metal cage

TTable 9.3 Notable Developments Relating to Implants

Year	Investigator	Development
Late 18th-19th century		Various metal devices to fix fractures; wires and pins from Fe, Au, Ag, and Pt
1860—1870	J. Lister	Asepitc surgical techniques
1886	H. Hansmann	Ni-plated steel fracture plate
1893-1912	W. A. Lane	Steel screws and plates for fracture fixation
1909	A. Lambotte	Brass, Al, Ag, and Cu plate
1912	Sherman	Vanadium steel plate, first alloy developed exclusively for medical use
1924	A. A. Zierold	Stellite® (CoCrMo alloy), a better material than Cu, Zn, steels, Mg, Fe, Ag, Au, and Al alloy
1926	M. Z. Lange	18-8sMo($2\% \sim 4\%$ Mo) stainless steel for greater corrosion resistance than 18-8 stainless steel
1926	E. W. Hey-Groves	Used carpenter's screw for femoral neck fracture
1931	M. N. Smith-Petersen	Designed first femoral neck fracture fixation nail made originally from stainless steel, later changed to Vitallium®
1936	C. S. Venable, W. G. Stuck	Vitallium: 19w/o Cr-9w/o Ni stainless steel
1938	P. Wiles	First total hip replacement
1946	J. and R. Judet	First biomechanically designed hip prosthesis; first plastics used in joint replacement
1940s	M. J. Dorzee, A. Franceschetti	Acrylics for corneal replacement
1947	J. Cotton	Ti and its alloys
1952	A. B. Voorhees, A. Jaretzta, A. H. Blackmore	First blood vessel replacement made of cltoth
1958	S. Furman, G. Robinson	First successful direct stimulation of heart
1958	J. Charnley	First use of acrylic bone cement in total hip replacements
1960	A. Starr, M. L. Edwards	Heart valve
1970s	W. J. Kolff	Experimental total heart replacement
1990s		Refined implants allowing bony ingrowth
1990s		Controversy over silicone mammary implants
2000s		Tissue engineering
2000s		Nanoscale materials

9.2 Nanotechnology and Nanomaterials

9.2.1 Introduction

In 1959 Nobel laureate physicist Richard Feynman delivered his now famous lecture, "There is Plenty of Room at the Bottom". He stimulated his audience with the vision of exciting new discoveries if one could fabricate materials and devices at the atomic/molecular scale. He pointed out that, for this to happen, a new class of miniaturized instrumentation would be needed to manipulate and measure the properties of these small- "nano" -structures.

It was not until the 1980s that instruments were invented with the capabilities Feynman envisioned. These instruments, including scanning tunneling microscopes, atomic force microscopes, and near-field microscopes, provide the "eyes" and "fingers" required for nanostructure measurement and manipulation.

What is nanotechnology?

Nanotechnology is ① the creation of useful materials, devices, and systems through the control of matter on the nanometer-length scale, and ② the exploitation of novel properties and phenomena developed at that scale.

What is a nanometer?

A nanometer is one billionth of a meter (10^{-9} m). This is roughly four times the diameter of an individual atom. A cube 2.5 nanometers on a side would contain about a thousand atoms. The smallest feature in an integrated circuit of today is 250 nanometers on a side and contains about one million atoms in a square layer of atomic height. Proteins, the molecules that catalyze chemical transformations in cells, are 1 to 20 nanometers in size. For comparison, a typical nanometer-scale feature size of about 10 nanometers is 1,000 times smaller than the diameter of a human hair.

Why is this length scale so important?

The wave-like (quantum mechanical) properties of electrons inside matter and atomic interactions are influenced by material variations on the nanometer scale. By creating nanometer-scale structures, it is possible to control fundamental properties of materials like their melting temperature, magnetic properties, charge capacity, and even their color, without changing the materials' chemical composition. Utilizing this potential will lead to new, high-performance products and technologies that were not possible before.

Systematic organization of matter on the nanometer length scale is a key feature of biological systems. Nanotechnology will allow us to place components and assemblies inside cells and to make new materials using the self-assembly methods of nature. In self-assembly, the information necessary for assembly is on the surface of the assembling nanocomponents. No robots or devices are needed to put the components together. This powerful combination of materials science and biotechnology will lead to entirely new processes and industries.

Nanoscale structures such as nanoparticles and nanolayers have very high surface-to-volume ratios, making them ideal for use in composite materials, chemical reactions, drug delivery, and energy storage. Nanostructured ceramics are often both harder and less brittle than the same materials made on the scale of microns, which are 1000 times larger than nanometers, but still just barely visible to the human eye. Nanoscale catalysis will increase the efficiency of chemical reactions and combustion, at the same time significantly reducing waste and pollution. More than half of therapeutically useful new medicines are not water soluble in the form of micron-scale particles, but they probably will dissolve in water if they are nanometer sized; thus nanostructuring greatly increases the chances of finding new drugs that can be rendered in usable forms.

Since nanostructures are so small, they can be used to build systems that contain a much higher density of components than micron-scale objects. Also, electrons will require much less time to move between components. Thus, new electronic device concepts, smaller and faster circuits, more sophisticated functions, and greatly reduced power consumption can all be achieved simultaneously by controlling nanostructure interactions and complexity.

These are just a few of the benefits and advantages of structuring materials at the nanometer scale.

Is this really new? Don't existing materials already use the nanometer-length scale?

Many existing technologies do already depend on nanoscale processes. Photography and catalysis are two examples of "old" nanotechnologies that were developed empirically in an earlier period despite their developers' limited abilities to probe and control matter at the nanoscale. These two technologies stand to be improved vastly as nanotechnology advances. Most currently existing technologies utilizing nanometer-scale objects were discovered by serendipity, and for many, the role that the nanometer scale played was not even appreciated until recently. For instance, we know now that adding certain inorganic clays to rubber dramatically improves the lifetime and wear properties of tires because the nanometer-sized clay particles bind to the ends of the polymer molecules, which are "molecular strings," and prevent them from unraveling. This is a simple process, but the dramatic improvement in the properties of this composite material, part rubber and part clay, demonstrates the great potential of nanotechnology as it is rationally applied to more complex systems. An example of such a system would be a structure designed to be extremely hard but not brittle, capable of self-repair if minor cracks appear, and easily broken down into its component parts when it is time to recycle the materials.

The ability to specifically analyze, organize, and control matter on many length scales simultaneously has only been possible for about the past ten years. For over a century, chemists have had the ability to control the arrangement of small numbers of atoms inside molecules, that is, to synthesize certain molecules with length scales of less than 1.5 nanometers. This has led to revolutions in drug design, plastics, and many other areas. Over the last several decades, photolithographic patterning (the primary manufacturing process of the semiconductor industry) of matter on the micron length scale has led to the revolution in microelectronics. With nanotechnology, it is just becoming possible to bridge the gap between atom/molecular length scale and microtechnology, and to control matter on every important length scale, enabling tremendous new power in materials design. It is important to remember that the most complex arrangements of matter known to us, living organisms, require specific patterning of matter on the molecular, nanometer, micron, millimeter, and meter scale all at once.

By tailoring the structure of materials at the nanoscale, it is possible to systematically and significantly change specific properties at larger scales — to engineer material behavior. Larger systems constructed of nanometer-scale components can have entirely new properties never before identified in nature. It is also possible to produce composites, i.e., mixtures of different nanoscale entities, that combine the most desirable properties of different materials to obtain characteristics that are greatly improved over those supplied by nature or that appear in combinations not produced by nature. Thus, nanotechnology encompasses a revolutionary set of principles, tools, and processes that will eventually become the foundation for such currently disparate applications as inks and dyes, protective coatings, medicines, electronics, energy storage and use, structural materials, and many others that we cannot even anticipate.

What will be the benefits of nanotechnology?

The new concepts of nanotechnology are so broad and pervasive that they may be expected to influence science and technology in ways that are unpredictable. We are just now seeing the tip of the iceberg in terms of the benefits that nanostructuring can bring (Figure 9.6). Existing products of nanotechnologies include wear-resistant tires made by combining nanometer-scale particles of inorganic clays with polymers; nanoparticle medicines with vastly improved delivery and control characteristics; greatly improved printing brought about by utilizing nanometer-scale particles with the best properties of both dyes and pigments; and vastly improved lasers and magnetic disk heads made by precisely controlling layer thicknesses. Many other applications are already under development or anticipated, including those listed below.

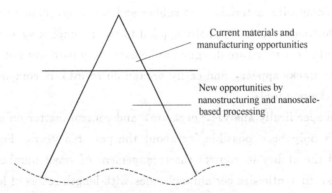

Figure 9.6 Current nanotechnology-related materials and manufacturing opportunities

- *Automotive and aeronautics industries*: nanoparticle-reinforced materials for lighter bodies, nanoparticle-reinforced tires that wear better and are recyclable, external painting that does not need washing, cheap non-flammable plastics, electronics for controls, self-repairing coatings and textiles.

- *Electronics and communications*: all-media recording using nanolayers and dots, flat panel displays, wireless technology, new devices and processes across the entire range of communication and information technologies, factors of thousands to millions improvements in both data storage capacity and processing speeds — and at lower cost and improved power efficiency compared to present electronic circuits.

• *Chemicals and materials*: catalysts that increase the energy efficiency of chemical plants and improve the combustion efficiency (thus lowering pollution emission) of motor vehicles, super-hard and tough (i.e., not brittle) drill bits and cutting tools, "smart" magnetic fluids for vacuum seals and lubricants.

• *Pharmaceuticals, healthcare, and life sciences*: new nanostructured drugs, gene and drug delivery systems targeted to specific sites in the body, biocompatible replacements for body parts and fluids, self-diagnostics for use in the home, sensors for labs-on-a-chip, material for bone and tissue regeneration.

• *Manufacturing*: precision engineering based on new generations of microscopes and measuring techniques, new processes and tools to manipulate matter at the atomic level, nanopowders that are sintered into bulk materials with special properties that may include sensors to detect incipient failures and actuators to repair problems, chemical-mechanical polishing with nanoparticles, self-assembling of structures from molecules, bio-inspired materials and biostructures.

• *Energy technologies*: new types of batteries, artificial photosynthesis for clean energy, quantum well solar cells, safe storage of hydrogen for use as a clean fuel, energy savings from using lighter materials and smaller circuits.

• *Space exploration*: lightweight space vehicles, economic energy generation and management, ultra-small and capable robotic systems.

• *Environment*: selective membranes that can filter contaminants or even salt from water, nanostructured traps for removing pollutants from industrial effluents, characterization of the effects of nanostructures in the environment, maintenance of industrial sustainability by significant reductions in materials and energy use, reduced sources of pollution, increased opportunities for recycling.

• *National security*: Detectors and detoxifiers of chemical and biological agents, dramatically more capable electronic circuits, hard nanostructured coatings and materials, camouflage materials, light and self-repairing textiles, blood replacement, miniaturized surveillance systems.

9.2.2 Examples of Current Achievements and Paradigm Shifts
Quantum Dot Formed by Self-assembly (Ge "pyramid")

Figure 9.7 is a scanning tunneling microscope (STM) image of a pyramid of germanium atoms on top of a silicon surface. The pyramid is ten nanometers across at the base, and it is actually only 1.5 nanometers tall (the height axis in the image has been stretched to make it easier to see the detail in the faces of the pyramid). Each round-looking object in the image is actually an individual germanium atom.

The pyramid forms itself in just a few seconds in a process called "self-assembly". If the proper number of germanium atoms is deposited onto the correct type of silicon surface, the interactions of the atoms with each other causes the pyramids to form spontaneously. The propensity of some materials to self-assemble into nanostructures is currently a major area of

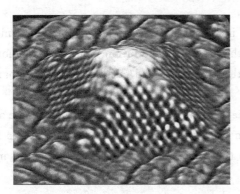

Figure 9.7 STM image of quantum dot formed by self-assembling (Ge "pyramid") (courtesy Hewlett-Packard; image acquired by G. Medeiros-Ribeiro, Hewlett-Packard Labs)

research. The intent is to learn how to guide or modify self-assembly to get materials to form more complex structures, such as electronic circuits. Manufacturing processes based on guided self-assembly of atoms, molecules, and supramolecules promise to be very inexpensive. Instead of requiring a multibillion-dollar manufacturing facility, electronic circuits of the future may be fabricated in a beaker using appropriate chemicals, and yet they may be many thousands to millions of times more capable than current chips. Just twenty years ago, few scientists even dreamed it would be possible to see such a detailed picture of the atomic world, but with the advent of new measuring tools, the scanning probe microscopes developed in the mid-1980s, seeing atoms is now an everyday occurrence in laboratories all over the world.

Quantum Corral

Figure 9.8 is a scanning tunneling microscope image of a "quantum corral." The corral is formed from 48 iron atoms, each of which was individually placed to form a circle with a 7.3 nanometer radius. The atoms were positioned with the tip of the tunneling microscope. The underlying material is pure copper. On this copper surface there are a group of electrons that are free to move about, forming a so-called "two-dimensional electron gas." When these electrons encounter an iron atom, they are partially reflected. The purpose of the corral is to try to trap, or "corral" some of the electrons into the circular structure, forcing the trapped electrons into "quantum" states. The circular undulations in the interior of the corral are a direct visualization of the spatial distribution of certain quantum states of the corral. Experiments such as this give scientists the ability to study the physics of nanometer-scale structures and to explore the potential application of these small structures to any of a number of purposes.

Figure 9.8 STM image of a "quantum corral" (courtesy IBM Research Division)

Magnetic Behavior at Nanoscale

Figure 9.9 is a transmission electron microscope (TEM) image of an elegant example of natural nanotechnology that occurs in magnetotactic bacteria. These are bacteria that contain within them a "compass" that allows them to move in a particular magnetic direction. The compass consists of a series of magnetic nanoparticles arranged in a line.

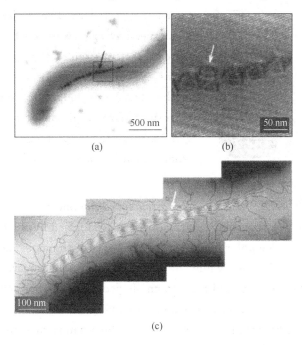

Figure 9.9 TEM images showing natural nanotechnology in magnetotactic bacteria (*magnetospirillum magnetotecticum* strain MS-1). Chains of nanocrystals used for navigation illustrate nature's exploitation of a fundamental scaling to achieve maximum and most efficient use of magnetization

Each particle is as large as it can be and still remain a single magnetic domain, 25nm. Larger particles have a type of defect, a magnetic domain wall, that lowers their coercivity. In the bacteria, these magnetic particles aggregate spontaneously into chains. The resulting compass uses a minimum amount of material to achieve the desired property, alignment along Earth's magnetic field. Artificial nanotechnology researchers can learn much from what already occurs in nature. The time required for a magnetized specimen to lose memory of its direction of magnetization depends exponentially upon the volume, provided the crystal is still in the nanometer regime. Thus, the same physics imposes a lower limit of a few tens of nanometers to the size of an iron oxide particle that could be used in a magnetic memory unit that operates at room temperature.

Ordered Monolayer of Gold Nanocrystals

Figure 9.10 is a transmission electron microscope image of an ordered monolayer of gold particles 5 nm in diameter supported on a thin carbon membrane. The round-looking objects are well-faceted single crystals of gold atoms. These nanocrystals have the shape shown in the bottom right inset and are aligned as a hexagonal array as illustrated by the expanded

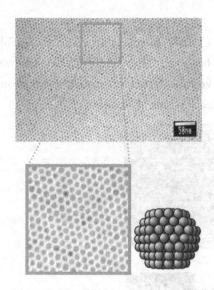

Figure 9.10 TEM image of an ordered monolayer of gold nanocrystals

view shown at the bottom left. Each gold particle is mechanically and electronically separated from its nearest neighbors by organic molecules, which give structural integrity to the monolayer and serve as a controlled tunnel barrier for electron transport between the particles. This ordered monolayer forms spontaneously on a water surface and can be transferred intact to a wide range of flat solid substrates. If instead of transferring the entire monolayer, a method can be developed to transfer only a selected pattern of narrow ribbons, it would provide an elegant solution to the problem of interconnecting electronic devices having smaller and smaller dimensions.

Nanostructured Polymers

Figure 9.11 depicts a supramolecular nanostructure formed by the ordered self-assembly of triblock copolymers. The polar liquid-crystalline parts of the molecules (bottom) arrange themselves in an ordered lattice, while the bulky, aromatic-hydrocarbon units (top) form an amorphous cap. These nanosized mushroom-shaped units further self-assemble into polar sheets whose top surfaces (mushroom caps) are hydrophobic and whose bottom surfaces (mushroom stems) are hydrophilic. Such self-assembled nanostructures are under investigation as anti-icing coatings for aircraft, lubricating layers for microelectronics, anti-thrombotic agents for arteries, etc.

Super-Strong Materials by Nanostructuring

Traditionally, the mechanical strength, s, of crystalline materials is believed to be largely controlled by the grain size d, often in the manner described by the Hall-Petch relationship, $\sigma = kd^{-\frac{1}{2}} + \sigma_0$. As the structural scale reduces to the nanometer range, researchers have found that the materials exhibit different scale dependence and there is a limit to the conventional descriptions of yielding (Misra et al. 1998). In addition to the high strength, the intrinsically high interface-to-volume ratio of the nanostructured materials may enhance interface-driven processes to extend the strain-to-failure and plasticity. A recent study on nanostructured Cu/Nb composites shows a complete suppression of brittle fracture when the wire was tensily tested at liquid He temperature (Han et al. 1998). This is an amazing finding, since bcc metals (such as Nb) are known to fracture in a brittle fashion at 4.2 K. The nanostructured Cu/Nb

Figure 9.11 Supramolecular nanostructure formed by the ordered self-assembly of triblock copolymers

composites exhibit significant strain hardening and ductility before fracture at a tensile strength of ~2 GPa and a strain of 10.

These results show that by reducing the structural scale to the nanometer range, one can extend the strength-ductility relationship beyond the current engineering materials limit, which is illustrated by the broad curve in the schematic diagram on the right side of Figure 9.12. A limitation of current engineering materials is that gain in strength is often offset by loss in ductility. The nanocomposite results suggest that by reducing the structural scale and by fully understanding the deformation physics governing the plasticity processes in nano-structured materials, we can produce materials with a combination of high strength and ductility (top right-hand corner of Figure 9.12).

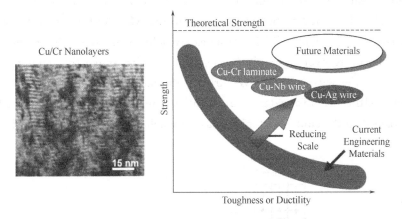

Figure 9.12 High resolution TEM image (left) of Cu/Cr nanolayers and diagram (right) of how reducing the scale of wire (materials) structure will affect the "toughness" and ductility of the materials. Nanostructured materials of the future will be able to transcend the limits of strength and ductility of current engineering materials

Due to their extremely complex nature and ultrafine structural scale, the characterization of deformation physics of nanostructured materials requires a close integration of state-of- the-art experimentation with atomistic modeling. Three-dimensional molecular dynamic (MD) simulations with up to 100 million atoms can now be performed with realistic interatomic potentials based on the embedded-atom method (Zhou et al. 1998; Holian and Lomdahl 1998). This opens up possibilities for unprecedented direct comparison between theory and experiments that will greatly enhance our understanding of the fundamental physics of materials with strength close to the theoretical limits.

Figure 9.13 Collection of nanoparticles of about 5 nm dia. (© 1999 Mark Reed. All rights reserved)

Three-Dimensional Structures for New Circuitry

Figure 9.13 depicts a collection of nanoparticles about 5 nm in diameter (gold particles in gold and semiconductor particles in green) with attached functionalized end groups (colored rods on particles) and buckyballs and nanotubes (black) on a substrate with gold electrodes. This is a rendition of a self-assembled molecular circuit produced by self assembly of the molecules and the ~5 nm gold nanoparticles using a thiol (S) chemistry process in a beaker.

9.3 Smart Materials

The shape memory effect (Figure 9.14) is observed when the temperature of a piece of shape memory alloy is cooled to below the temperature M_f. At this stage the alloy is completely composed of Martensite which can be easily deformed. After distorting the SMA the original shape can be recovered simply by heating the wire above the temperature A_f. The heat transferred to the wire is the power driving the molecular rearrangement of the alloy, similar to heat melting ice into water, but the alloy remains solid. The deformed Martensite is now transformed to the cubic Austenite phase, which is configured in the original shape of the wire.

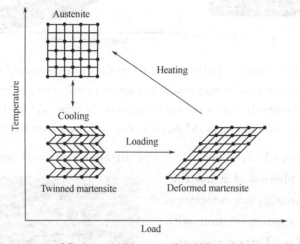

Figure 9.14 Microscopic Diagram of the Shape Memory Effect

9.3.1 Introduction

Smart materials are materials that have one or more properties that can be significantly changed in a controlled fashion by external stimuli, such as stress, temperature, moisture, pH, electric or magnetic fields.

There are a number of types of smart material, some of which are already common. Some examples are as following:

• Piezoelectric materials are materials that produce a voltage when stress is applied. Since this effect also applies in the reverse manner, a voltage across the sample will produce stress within the sample. Suitably designed structures made from these materials can therefore be made that bend, expand or contract when a voltage is applied.

• Shape memory alloys and shape memory polymers are materials in which large deformation can be induced and recovered through temperature changes or stress changes (pseudoelasticity). The large deformation results due to martensitic phase change.

• Magnetostrictive materials exhibit change in shape under the influence of magnetic field and also exhibit change in their magnetization under the influence of mechanical stress.

• Magnetic shape memory alloys are materials that change their shape in response to a significant change in the magnetic field.

• pH-sensitive polymers are materials which swell/collapse when the pH of the surrounding media changes.

• Temperature-responsive polymers are materials which undergo changes upon temperature.

• Halochromic materials are commonly used materials that change their colour as a result of changing acidity. One suggested application is for paints that can change colour to indicate corrosion in the metal underneath them.

• Chromogenic systems change colour in response to electrical, optical or thermal changes. These include electrochromic materials, which change their colour or opacity on the application of a voltage (e.g. liquid crystal displays), thermochromic materials change in color depending on their temperature, and photochromic materials, which change colour in response to light—for example, light sensitive sunglasses that darken when exposed to bright sunlight.

• Another good example is custard, as long as it is starch-based.

• Ferrofluid

• Photomechanical materials change shape under exposure to light.

• Self-healing materials have the intrinsic ability to repair damage due to normal usage, thus expanding the material's lifetime.

9.3.2 Shape Memory Alloys

With the shape memory effect, heating reverses prior plastic deformation. Alloys that exhibit this effect are ordered solid solutions that undergo a martensitic transformation on cooling.

Martensitic transformations occur by nucleation and shear. There is no composition change and hence no growth by diffusion. Rather, a region of the lattice suddenly transforms by shear. The elastic energy caused by misfit of the new and old lattice is minimized if the region undergoing the transformation is lenticular. Figure 9.15 shows the effect of a shear strain of $\gamma=2$ on a spherical particle (top) and an ellipsoidal particle of the same volume (bottom). With the ellipsoidal particle, there is much less disturbance of the surrounding matrix. A martensitic transformation occurs over a temperature range. The temperature at which the martensite first starts to form on cooling is called the M_s temperature. More martensite will form only if the temperature is lowered. The temperature at which the reaction is complete is called the M_f temperature. However, the concept of an M_f temperature may be more of a convenience than a reality because often there is no sharp completion of martensite formation.

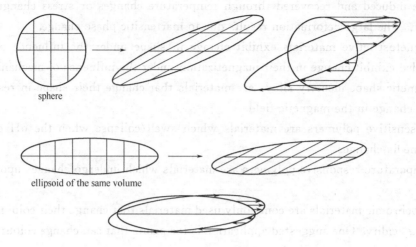

Figure 9.15 A spherical region undergoing a martensitic shear distorts a much larger volume of material than an ellipsoid of the same volume

In most systems the martensitic reaction is geometrically reversible. On heating, the martensite will start to form the higher temperature phase at the A_s temperature and the reaction will be complete at an A_f temperature, as illustrated in Figure 9.16. Martensite in the iron—carbon system is an exception. On heating, the iron—carbon martensite decomposes into iron carbide and ferrite before the A_s temperature is reached. Martensite can be induced to form at temperatures somewhat above the M_s by deformation. The highest temperature at which this can occur is called the M_d temperature. Likewise, the reverse transformation can be induced by deformation at the A_d temperature somewhat below the A_s. The temperature at which the two phases are thermodynamically in equilibrium must lie between the A_d and M_d temperatures.

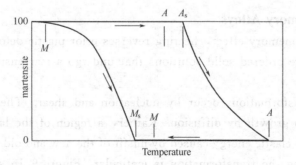

Figure 9.16 The fraction martensite increases as the temperature is lowered below the M_s to the M_f. On heating, the reversion starts at A_s and finishes at A_f

The volumes of the low and high temperature phases are usually not the same so there is often a volume change associated with the martensite reaction.

Shape memory effects were first observed in AuCd in 1932, but it was the discovery by

Buehler et al. in of the effect in NiTi in 1962 that stimulated interest in shape memory. The alloy TiNi (49 to 51 atomic % Ni) has an ordered bcc structure at 200℃. On cooling it transforms to a monoclinic structure by a martensitic shear. The shear strain associated with this transformation is about 12%. There is more than one variant of the transformation. If only one variant of the martensite were formed, the strain in the neighboring untransformed lattice would be far too high to accommodate. Instead two mirror image variants form in such a way that there is no macroscopic strain. The macroscopic shape is the same as before the transformation. Figure 9.17 illustrates this. The boundaries between the two variants are highly mobile. The resulting structure can deform easily by movement of these boundaries. Figure 9.18 shows stress—strain curves above and below the M_s. Heating the deformed material above the A_f temperature causes it to transform back to the ordered cubic structure by martensitic shear. The overall effect is that the deformation imposed on the low temperature martensitic form is reversed on heating. The critical temperatures for reversal in TiNi alloys are typically in the range of 80 to 100℃ but are sensitive to very minor changes in composition so material can be produced with specific reversal temperatures. Excess nickel greatly lowers the transformation temperature. It is also depressed by small additions of iron and chromium. Copper decreases the hysteresis.

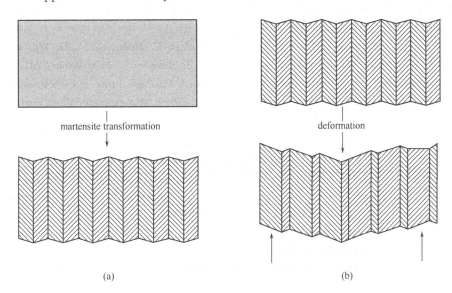

Figure 9.17 (a) As the material is cooled it undergoes a martensitic transformation. By transforming to equal amounts of two variants, the macroscopic shape is retained. (b) Deformation occurs by movement of variant boundaries so the more favorably oriented variant grows at the expense of the other. Reprinted with permission of Cambridge University Press from W. F. Hosford, *Mechanical Behavior of Materials* (New York: Cambridge Univ. Press, 2005)

Other shape memory alloys are listed in Table 9.4, and Figure 9.19 shows the dependence of the M_s temperature for Cu-Zn-Al alloys on composition. For the copper-base alloys controlled cooling is necessary after heating into the β phase region.

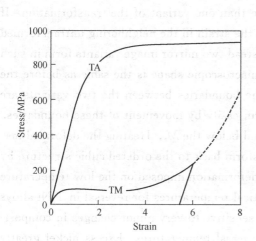 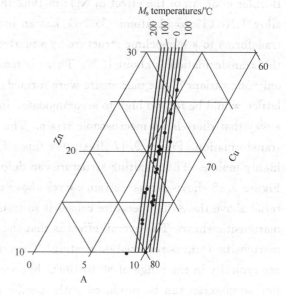

Figure 9.18 Stress-strain curve for a shape memory material. The lower curve is for deformation when the material is entirely martensitic. The deformation occurs by movement of variant boundaries. After all of the material is of one variant, the stress rises rapidly. The upper curve is for the material above its A_f temperature. (Adapted from a sketch by D. Grummon.)

Figure 9.19 The dependence of the M_s temperature of Copper-Zinc-Aluminum alloys on composition. The dots indicate alloys for which the M_s has been measured. (Data from D. E. Hodgson, M. H. Wu, and R. J. Biermann, *Shape Memory Alloys*, Johnson Matthey, http://www.jmmedical.com/html/shape memory alloys html.)

Table 9.4 Alloys with shape memory

Alloy	Composition	Transformation temperature range/℃	Temprature hysteresis/℃
AuCd	46.5%～50%at. Cd	30～100	15
Cu-Al-Ni	14%～14.5wt. Al 3%～4.5%wt. Ni	−140～100	35
Cu-Zn-X X = a few wt. % Si, Sn or Al.	38.5%～41.5%wt. Zn	−180～200	10
In-Ti	18%～23%at. Ti	60～100	4
Ni-Al	36%～38%at. Al	−180～100	10
NiTi	49%～51%at. Ni	−50～110	30

Shape memory in polymers

When a polymer is deformed at temperatures below its glass transition and then heated above the glass transition temperature, its shape will revert to that it originally had before being deformed. The amount of reversible strain is much larger than in metals (up to 400% for polymers vs. less than 10% for metals). This effect is utilized in shrink-wrapping of consumer products. Films are stretched biaxially at temperatures below their glass transition temperature. After the product is wrapped, the temperature is raised above the glass transition temperature by warm air, allowing the film to shrink tightly around the product. This effect is also used for insulating wiring joints. Preexpanded tubes are slipped over the joints. Heating causes them to shrink tightly around the bare wires.

Medical applications of biodegradable shape memory polymers include their use for removing blood clots formed during strokes. Preshaped foams can be used to fill cranial aneurisms. Loosely tied sutures made from fibers that have been stretched at 50℃ will tighten when heated just above room temperature.

9.3.3 Applications of Smart Materials

There are many possibilities for such materials and structures in the man made world. Engineering structures could operate at the very limit of their performance envelopes and to their structural limits without fear of exceeding either. These structures could also give maintenance engineers a full report on performance history, as well as the location of defects, whilst having the ability to counteract unwanted or potentially dangerous conditions such as excessive vibration, and effect self repair. The Office of Science and Technology Foresight Programme has stated that "Smart materials... will have an increasing range of applications (and) the underlying sciences in this area... must be maintained at a standard which helps achieve technological objectives", which means that smart materials and structures must solve engineering problems with hitherto unachievable efficiency, and provide an opportunity for new wealth creating products.

Smart Materials in Aerospace

Some materials and structures can be termed 'sensual' devices. These are structures that can sense their environment and generate data for use in health and usage monitoring systems (HUMS). To date the most well established application of HUMS are in the field of aerospace, in areas such as aircraft checking.

An airline such as British Airways requires over 1000 employees to service their 747s with extensive routine, ramp, intermediate and major checks to monitor the health and usage of the fleet. Routine checks involve literally dozens of tasks carried out under approximately 12 pages of densely typed check headings. Ramp checks increase in thoroughness every 10 days to 1 month, hanger checks occur every 3 months, "interchecks" every 15 months, and major checks every 24000 flying hours. In addition to the manpower resources, hanger checks require the aircraft to be out of service for 24 hours, interchecks require 10 days and major checks 5 weeks. The overheads of such safety monitoring are enormous.

An aircraft constructed from a 'sensual structure' could self-monitor its performance to a level beyond that of current data recording, and provide ground crews with enhanced health and usage monitoring. This would minimize the overheads associated with HUMS and allow such aircraft to fly for more hours before human intervention is required.

Smart Materials in Civil Engineering Applications

However, 'sensual structures' need not be restricted to hi-tech applications such as aircraft. They could be used in the monitoring of civil engineering structures to assess durability. Monitoring of the current and long term behavior of a bridge would lead to enhanced safety during its life since it would provide early warning of structural problems at a stage where minor repairs would enhance durability, and when used in conjunction with structural rehabilitation could be used to safety monitor the structure beyond its original design life. This would influence the life costs of such structures by reducing upfront construction costs (since smart structures would allow reduced safety factors in initial design), and by extending the safe life of the structure. 'sensual' materials and structures also have a wide range of potential domestic applications, as in food packaging for monitoring safe storage and cooking.

The above examples address only 'sensual' structures. However, smart materials and structures offer the possibility of structures which not only sense but also adapt to their environment. Such adaptive materials and structures benefit from the sensual aspects highlighted earlier, but in addition have the capability to move, vibrate, and exhibit a multitude of other real time responses.

Potential applications of such adaptive materials and structures range from the ability to control the aeroelastic form of an aircraft wing, thus minimizing drag and improving operational efficiency, to vibration control of lightweight structures such as satellites, and power pick-up pantographs on trains. The domestic environment is also a potential market for such materials and structures, with the possibility of touch sensitive materials for seating, domestic appliances, and other products. These concepts may seem 'blue sky', but some may be nearing commercial readiness as you read this.

The Future

The development of true smart materials at the atomic scale is still some way off, although the enabling technologies are under development. These require novel aspects of nanotechnology (technologies associated with materials and processes at the nanometer scale, 10^{-9} m) and the newly developing science of shape chemistry.

Worldwide, considerable effort is being deployed to develop smart materials and structures. The technological benefits of such systems have begun to be identified and, demonstrators are under construction for a wide range of applications from space and aerospace, to civil engineering and domestic products. In many of these applications, the cost benefit analyses of such systems have yet to be fully demonstrated.

Reference

[1] Hosford W. F., *Materials Science*: INTERMEDIATE TEXT, 2007, Cambridge University Press, The Edinburgh Building, Cambridge cb2 2ru, UK.